E

"*Ephraim's Farm* is a historical sketch of a western Pennsylvania working farm family of the early 1900s. Although this family may be ordinary in every detail, the totality of the lives so factually described is extraordinary. The Romesbergs represent a single but strong thread in the fabric that was rural America. This American story of family, God, self-reliance and hard, physical work is now a memory and will not be repeated, but should never be forgotten. It should be required reading for our children so that they would have the opportunity to absorb even the smallest morsels of the abundant integrity that this family unknowingly and innocently displayed."

—Irvin Pritts, Ph.D.

"Although monetarily poor, the large family that the author was born into and writes about was truly blessed with family values and a caring concern for each other. The author shares many seemingly insignificant stories that help portray the underlying love that existed between the members of the family. The book preserves many daily farming activities and styles of country living in southwestern Pennsylvania from a time period that has completely changed. As you proceed through the book, the stories have a way of building one's desire to visit the farming and coal mining community where the Romesberg farm was located. *Ephraim's Farm* keeps your interest with stories about Henry Pritts' trial, eventually being hung, and then buried in a lone grave near the border of the Romesberg farm. The ghost stories associated with Pritts' grave add extra suspense."

—David R. Hay

"What an intriguing history of the Romesberg family! Once you begin to read this history you will want to continue reading on to the very end."

—Betty Arnold

"This true account of the Romesberg family is so factual it transported me back to my childhood—which was about a mile away—in Wison Creek."

—Avis Engleka, cousin

"I am a friend of Floyd Romesberg from his days living in Midland, Michigan. We hunted deer together, attended the same church, are both engineers, past members of Toastmaster's International, and I am living on my first farm. This book was intensely interesting to me. I read it carefully to absorb details and add understanding to stories from my own family as I grew up. Organization of the book is excellent with chronological information starting with immigration, through today's life. There are many fine stories, a message from the sky, description of farm living and then memoirs added by family members with yet new stories and different info on previous ones. Many of the stories could have been describing my own family. I would like to have a book like this written for my family. I suppose I will have to write and edit that book. My grandmother's family were Quakers from Plymouth Colony in 1635 and from Pennsylvania Colony in 1683. My grandfather's family are descendents of German immigrants who homesteaded near Tiffin, Ohio in 1830. We still attend reunions there every year. *Ephriam's Farm* renewed many memories and helped me to understand my roots."

—Dennis Carey Zeiss

Ephraim's Farm

A Memoir of
Rural Pennsylvania

Floyd E. Romesberg

BLUE DOLPHIN PUBLISHING

Published by Blue Dolphin Publishing, Inc.
P.O. Box 8, Nevada City, CA 95959
Orders: 1-800-643-0765
Web: www.bluedolphinpublishing.com

ISBN: 978-1-57733-249-0

Library of Congress Cataloging-in-Publication Data

Romesberg, Floyd E., 1927-
    Ephraim's farm : a memoir of rural Pennsylvania / Floyd E. Romesberg.
       p. cm.
    Includes bibliographical references.
    ISBN 978-1-57733-249-7 (alk. paper)
    1. Romesberg, Floyd E., 1927- 2. Romesberg, Floyd E., 1927—Child-
hood and youth. 3. Romesberg, Floyd E., 1927—Family. 4. Romesberg,
Ephraim J., 1878-1964—Family. 5. Farms—Pennsylvania—Black
(Township)—History. 6. Farm life—Pennsylvania—Black (Township)—
History. 7. Black (Pa. : Township)—Social life and customs. 8. Black (Pa.
: Township)—Biography. 9. Black (Pa. : Township)—Genealogy. I. Title.
    F159.B58R66 2010
    929'.20973—dc22
                                                            2009049660

Printed in the United States of America

10   9   8   7   6   5   4   3   2   1

Contents

Preface

Our family name evolved from a long process of choice. Names were made up from a list of the following[1]:

- ➤ Rams, Roms, Ranes, Rems, Romes, Rimens, Ramens, Ramps,
- ➤ burgh, barger, burg, berg, perger, and burger.

Many combinations were possible:

- ➤ Riemensperger
- ➤ Rimensperger
- ➤ Rimensberger
- ➤ Rymensoerger
- ➤ Remsberger
- ➤ Ramenspery
- ➤ Ramsperger
- ➤ Romesburg
- ➤ Romesberg

Some were rich. One built a castle and another bought a castle. They controlled land and water. Others were farmers. They were masters of the soil, which they tilled with wooden hoe and shovel.

The well-to-do families usually placed their name on their own Coat of Arms. Usually the richest of the families had armies to protect their land and name.

Family history of these early German families is somewhat limited. Even today when some people die, history is lost because events are not recorded. More and better-organized information should be recorded through pictures and writings. It is hoped that this writing deals less with the roots of our ancestors and more about the events that surrounded their lives.

Coat of Arms[1]

ROMESBERG

Certificate of Authenticity [1]

This is to certify that the Coat of Arms described hereon has been used in centuries past by a person or family with the Surname:

ROMESBERG

or an onomatological variant thereof and is therefore judged to be associated with this name.

Arms Description:
"Per fess; 1st argent a trefoil gules pierced of the field, 2nd argent a ship of 3 masts adorned with banners gules, carrying an imperial pavillion sailing in the open sea, all proper."

Reference Source:
Les Planches de l'Armorial General
by T. P. Rietstap

Manuscript Number: 030671-06

Confirmed by the
Presiding Officer of the
SANSON INSTITUTE
of HERALDRY

Prayer,
2006 Ephraim Romesberg Family Reunion

A Prayer of Thanks and Praise

Almighty God, we come to you in the name of Jesus Christ, the author and finisher of our faith. Heavenly Father, I come to you in behalf of the family. We ask that your hand of protection be upon each one, and that our lives would be an example to those we come in contact with and would enlighten their lives here on this earth. Then we would enjoy life in itself. You also promised us we should prosper and be in good health.

Help each one of us to lean upon you for our every need. We have no one else to turn to. You hold life and death in your hands. Help us to be ready when our time comes to leave this world. Help each one of us to cast our cares upon you that we can hear you say, "Enter into the joys of the Lord."

God Bless each one of you in the Name of The Lord Jesus Christ, our Savior.

Love, Brother Paul

The Old Homestead

The house as pictured on the front cover had its beginning in about 1790 as a log cabin built by the Pritts family.

I can envision the Pritts family, probably working from daylight to dark. Early morning usually brought the sound of a soft breeze blowing through the leaves of a stand of red oak trees. Then there might be the calls of mourning doves from near to far. The rasping sound of the men's saws went on and on. The chopping sound of the adze and broad axe accompanied the flight of wood chips from the log.

Now and then someone would lay down his tool and walk about twenty feet to fill his cup with the cold spring water. These hard working men are long gone, but the faint rippling sound of this little stream has continued without stopping from that day until the present.

The old homestead was purchased by Homer and Jean Colflesh in 1967. The property is now owned by their son, Jeffery, who was a Hollywood movie script writer. Jeffery was largely responsible for the fourth and splendid remodeling and modernizing of our old home, both inside and out.

Acknowledgments

My sincere thanks to all those who helped and supported me in the writing of this book.

Irvin Pritts and Avis Engleka, who provided information on the Pritts family.

Brothers Paul and E. Jay, sisters Luella, Betty, Elaine, nephew John Ogle and my niece JoAnn Caruso for their memoirs.
Also my cousin, Minnie Thomas, for the biography of her father, Uncle Milt.

My nephew, Merle Romesberg, Jr. and his wife, Hazel, for providing family information.

Betty Arnold of the Rockwood Area Historical and Genealogical Historical Society for some of the early photographs.

Blanche Pederzelli (my first cousin) and Aunt Della's daughter, who had provided much information for "The Romesberg Family History."

Austin Romesberg, who provided early genealogy information about European Romesbergs to my sister, Betty Romesberg Clark, who lives in Switzerland.

A very special thanks to my nephew, John Ogle, for all the encouragement and support he has given me this past year in my efforts to finish my book. He has typed most of the manuscript, helped with the organization of the material, as well as communicating with me on the telephone very frequently. He has been most generous with his time!

My brother, E. Jay, and my wife's granddaughter, Katie Hilt, who also did some of the typing.

Carol Ogle, John's wife, who provided most of the illustrations. She is an accomplished artist and I am most appreciative.

John and our young friend Emily Morton also contributed several illustrations.

Katie Goundy, my daughter Beverly's daughter, who is a graphic design student, designed the book cover.

Shirley, my wife, many thanks for proofreading and correcting grammar.

Introduction

Immigrants came to America from many countries. They played a key roll in the growth of this great country. The promise of free land and a better life filled them with hope.

They brought many skills and a broad band of knowledge from each of their mother countries. These early settlers all faced similar problems. The land was covered with huge trees, which would need to be cut down and their stumps removed before homes could be built and crops planted. The forest held an abundance of wild game. Many became hunters to supply meat for the table. Many became trappers and sold the furs to buy needed items. Lumbermen supplied lumber and carpenters built houses.

Stone masons built both homes and foundations using the many stones that were scattered over the land. Many types of stump pullers were developed to clear the land for homes and planting of crops. Wagons, tools and other machinery were needed and the blacksmith, wheelwrights, and welding trades were developed.

Many small towns appeared, surrounded by rural communities. Some of these once small towns grew into large cities and many rural people moved to urban areas in search of work as time went on. As many moved away from the farms, families became smaller. The days of big family gatherings on Sunday for dinner and maybe a ball game in the afternoon became a thing of the past.

As we look a little closer at our American culture, we can see people scurrying about, everyone going in different directions, doing different things, everyone with a different agenda. For many, family life as we had known it began to disappear. With the invention of TV we have seen many changes in our society. TV has provided entertainment in the homes, which was enjoyable for the most part, but with the increasing display of violence such as robberies, murders, drugs and sexually explicit information, many of our values have changed. Sadly, our children have been exposed to all this ugliness.

What can we do about all this?

Perhaps you will find some answers in the pages of *Ephraim's Farm* as you read the story of this family. The word "love" is not mentioned, yet "acts of love" were shown continuously throughout the story.

"If a young boy plays baseball every Sunday afternoon with his brothers, you will not find him on the street with drugs in one pocket and a gun in the other."

It is clear that we need to bring back "family," strong parenting, and training of youth, starting at a young age.

Chapter 1

Escape to America

Sketch by Carol Ogle

Ship – to America

Family folklore has been passed from generation to generation regarding the reasons for the extensive immigration to America. One such story claims a native German family returned to their homeland after the Thirty Years' War to find the land covered with dust, the trees mangled and charred, and many stones scattered everywhere. Some were arranged in square shapes depicting where a house once stood. Some, as far as the eye could see, stood in silence marking a grave of a fallen soldier, or a father, mother, or child. Closer observation showed bones everywhere, bones of cattle, horses, sheep, and dogs. The land gave up a faint odor of decaying flesh.

Many joined hands and declared, "This tragedy must be laid to rest behind us. With hard work and determination we will build a bright future in America."

A major factor in the flow of immigrants from Europe to America was the Thirty Years' War,[2] which lasted from 1618 to 1648.[3] The impact was devastating and lasted for many years after the war ended. The war began with a revolt in Bohemia and spread to Germany. It involved Britain, Spain, the Netherlands, Denmark, Sweden, and France. A large part of the war was fought on German ground. At this time, Germany was divided into three hundred principalities, or governing bodies, which were further divided among religious lines—Catholicism, Lutheranism, and Calvinism.

As a result of Germany's disorganization during the war, it was demolished. Many died from battle, and many from disease and starvation. As many women and children died as men. There were many homeless and many orphans.

Flight to America became a norm. Some European businessmen promoted the immigration by getting the colonies to accept a wide range of people who wanted to come to America. This especially included people without homes or jobs. People were taken from jails. There were people from broken families and many orphans. Groups of people were loaded on ships. Tickets were given to many who couldn't afford to buy their own. Many strangers met for the first time on the ships during the long journey across the Atlantic to America.

William Penn became a giant in the immigration story. Penn acquired Pennsylvania from the king of England in payment for a debt, which the king owed his father when Penn's father died[4].

Many German people had two choices: Come to America, the land of opportunity, or stay in a battle-torn land, plagued with sickness and disease. William Penn painted a picture of Pennsylvania: a land of tall oak and maple trees; bubbling, trout-filled mountain streams; wild turkey, deer, and small game; fertile valleys between mountains with straight rows of corn growing alongside of blackberries and blueberries beckoning to be picked. Pennsylvania was a land of opportunity.

Our Romesberg ancestors lived through the epic efforts of the war and this fateful event brought us to America.

FROM PAST TO PRESENT[5]

Hans Rymensperger – Born 1615 in Switzerland; Wife Anna Keller

Jakob Rymensperger – Born 1640 in Switzerland; Wife Anna Bischoff

Jakob Wolfgang Riemensperger – Born 1670 in Germany; Died 1712; Wife Helena Elizabette

Johann Stephen Riemensperger – Born 1711 in Germany; Died 1789 in MD; Wife Anna Catharine Brunner from Germany to PA to MD.

Elias Romesburg – Born 1748 in MD; Died 1803 in PA; Wife Catharine Schaffer (or Houck?)

Stephen Romesburg – Born 1782; Died 1848 in PA; Wife Maria

Solomon Romesburg – Born 1812; Died in Black Twp.; Wife Polly (Mary) Pritts

Levi Romesburg – Born 1840; Died 1889 in Black Twp.; Wife Mary Ann Livingston

Ephraim J. Romesberg – Born 1878; Died 1964 in Black Twp.; Wife Susan Mayme Swearman: children; Mary, Luella, Merle, Wilbur, Della, Helen, Paul, Betty, Elaine, Floyd, Ephraim

Chapter 2

Stephen and Elias Lead the Way

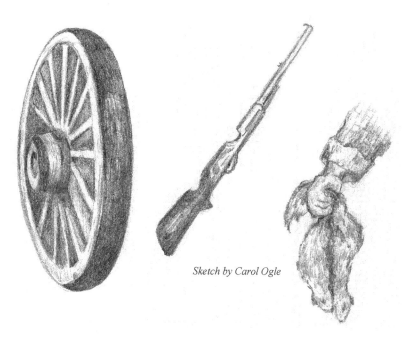

Sketch by Carol Ogle

Our early ancestors were born in and lived in Germany. Hans was born three years before the start of the Thirty Years' War. His wife Anna Keller and their six children lived in the small town of Bush, Switzerland. His son, Jacob, and his grandson, Jacob W., lived in the early years after the war. Jacob W. lived during William Penn's stay in America. The miraculous event that brought Stephen to America in 1719 at 8 years of age is not known. Records show he was in Lancaster, Pennsylvania. He may have come along with a group of orphans. Most likely his mother came with him on the boat. Many died on the long trip from disease and poor diet.

The story of Stephen's life in Europe and in America is one of remarkable success, a story of rags to riches. Likely, he was taken in by a family. He received a good education and was married at twenty-nine to Maria Catharine Brunner. The Brunner family had recently emigrated from Germany. He was a captain in the French and Indian War. He was a very successful farmer, wheelwright, and salesman. He bought and sold land, both farm land and timber. His Last Will and Testament indicated that he had accumulated wealth by the time of his death. He had four Afro-Americans working on his farm. He included them in his will to make sure of their freedom and financial security. In his will, he showed about $2,000 and 300 acres of land going to three of his seven children with the remainder divided among all seven.

In his lifetime, Stephen was looked upon as a man among men. He developed a community and church and was a leader in both.

There is no way to tell from his will his total worth at his death. Indications are that he did well at whatever he undertook. His choice of a trade (wheelwright) was a sound business decision because of the growing need for wagon wheels for the many uses on farms and in towns.

Stephen's descendants, Elias and Stephen, were responsible for moving the family to Somerset County and the little towns of Blackfield and Wilsoncreek. Elias used the term "plantation" in his will. The Romesberg Genealogy claims, at his death, that he was worth $1,200,000 and 2,000 acres of land. Elias was recognized as a great hunter. There is a story telling that he led a group of hunters who shot five bear the first year in Somerset County.

Chapter 3

The Pritts Family

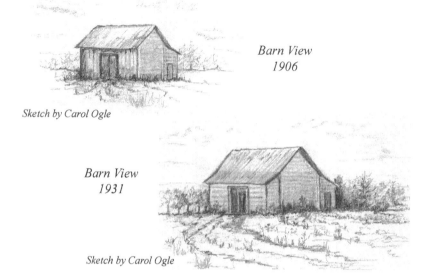

Barn View
1906

Sketch by Carol Ogle

Barn View
1931

Sketch by Carol Ogle

The early pioneers who settled in the hills of Pennsylvania had the formidable task of building a log cabin, a barn and clearing land to grow crops.

Jacob Pritts[6] was born about 1744 in Germany. He immigrated to the Philadelphia area with his seven-month pregnant wife and two young brothers, George and Samuel. They arrived September 19, 1765 on the ship "Betsey."

Jacob had three sons, George (b. 6/17/1769), Jacob II (b. 6/17/1780) and John (b about 1793). Also, there were two daughters, Elizabeth (b. 1781) and Mary (b. 1791).

Jacob's son, George, blazed the trail for the rest of the family by moving to western Milford Township, Somerset County. Jacob remained in the Philadelphia area for thirty-two years. He then moved to

Milford Township in an area know as Wilsoncreek. Jacob died about 1807. Jacob was a carpenter and likely played a key role in building the cabins, barn, and house. During the first ten to twenty years, it would appear that there may have been as many as ten adults available to work. A good spring supplied a heavy flow of fresh water. It is believed that two cabins were built close to the barn and one across the road where Uncle Milt's house was later built.

One of the original cabins was much larger than the others. The sides of the house consisted of 4" x 14" sawed logs placed vertically on all sides.

Sometime later, this cabin was converted into a frame house. It was modified and enlarged in 1920 by Uncle Joe and my Dad, Ephraim.

The barn was apparently built in three steps. First, two large rooms about 25 feet x 25 feet x 18 feet high were constructed out of logs. These rooms were built 25 feet apart to make a third room, which was called a threshing room or floor. A roof, 75 feet long, covered the barns. Horses and wagons pulled loads of hay through the big side doors and the hay was unloaded into the rooms on either side.

Three stanchions for horses were built in one of the rooms. Hay was stored on the top floor built above the stanchions.

Three horses was typical for many farmers. Two horses were used for many jobs such as plowing, harrowing, cultivating, pulling wagons, planting crops, etc. Three horses were best for some jobs, such as combining grain. One horse was used to pull the buggy to go to church, to town and to do light work such as raking hay.

In the second step, a shed (about 25 feet by 12 feet) was built along the back side. This area was used for milk cows and young cattle.

The third step was completed in 1931. A shed was built along the front side. The center room was an extension of the threshing floor. The two outside rooms were used for storage of hay.

It is not known how much land was already cleared when the Pritts family first arrived. About 85 acres were being plowed by 1900. Removing stumps was very difficult. They likely used stump pullers and horses. This was a dangerous task. Uncle Ralph had a serious accident when the hitch broke and the beam swung back and hit him on the leg. He lost the leg.

Jacob II and John, with their families, stayed on or near the farm in Wilsoncreek. John died at age 37 in1830, leaving three sons and six daughters and his wife. Jacob II raised some of John's family. Jacob II had a son Israel, a son Henry (b. 1800), and a daughter Polly (b. 1802).

Around 1860, Polly Pritts, who had married Solomon Romesberg, inherited the farm. Their son, Levi was my grandfather.

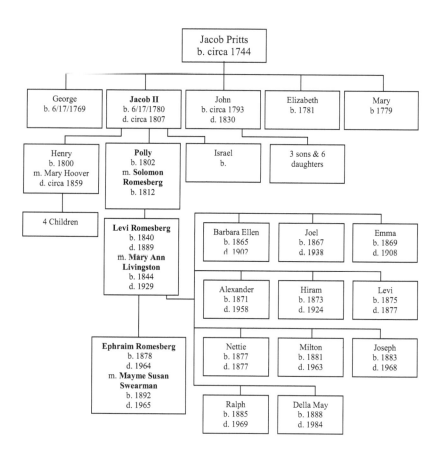

The first-born son (1800) of Jacob II was Henry. Henry played an important role in the development of our farm near Wilsoncreek. Henry was the third great uncle of Irvin Pritts who wrote the following:

"When Henry was 21, Jacob sold him two mares for $80, four cows and one calf for $50, seven sheep for $41, six hogs for $16, one windmill for $16, and one stove for $10, one wagon swing latres and steerers for $25. Likewise, all household and kitchen furniture for $50, and forty acres of winter grain for $200. Even by today's standards the 21-year-old Henry seemed to have a good start in life. He was married in his late thirties to Mary Hoover and had four children."[7]

The following article was taken from the Somerset Bicentennial (1795-1995) publication "Generation Upon Generation":[8]

Henry Pritts and Eli Weimer, both of Milford Township, went hunting together in October of 1859. Pritts returned and said that the two men had separated, and he didn't know where Weimer was. Weimer's body was found a week later. It appeared that he had committed suicide, but Pritts was arrested for murder when it was learned he was enamored of Weimer's wife. Judge Francis M. Kimmell presided over the trial. Pritts was convicted, and admitted his guilt shortly before he was hanged.

His friends buried his body on the farm where he lived. This farm later belonged to Milt Romesberg. The gravesite was located about seven feet from the fence line and about twenty feet from the wood line on Ephraim Romesberg's farm. A small stone marked the grave. Our family plowed around this stone for about forty years. Around 1940, Uncle Alex told my brother Merle to take the stone away to make farming a little easier. Merle told me all the same, so I carried and rolled the stone about seven feet to the fence.

I was introduced to some of the Pritts family when I was ten years old, in 1937. I learned a little about the skills of several of the family. My brother Merle had decided to start a baseball team. The first thing he needed was a bat. To this end he cut a piece of white ash, about 3½ feet long and 8 inches in diameter. He put it in the barn to dry. He invited me to go along to see Israel Pritts. Merle carried the piece of ash to the neighbor's house in Wilsoncreek and handed it to the old man. He asked him if he would make a bat for us. He looked old to me—like 100. However, he politely agreed and then invited us to the ball game. Merle was happy to be invited and remarked to me, "Let's go see Luther play."

Luther was Israel's oldest son. Merle said, "Watch Luther catch the ball." It took me a while to see what he did. With a glove in his left hand, he reached out for the ball with the back of the glove facing the ball in flight.

"Crack!"

The ball hit his glove. I said, "He missed it!"

But then he reached up with his right hand and pulled the ball from his glove and threw it to the first baseman. Merle explained, "He reached for the ball with the back of the glove facing the oncoming ball, then flipped his hand so quickly I couldn't tell he flipped it."

Then Luther came to bat. He hit the ball into the woods in left field. He ran around the bases and when he reached third base, he slowed up and walked home on his hands! Wow, what a ball player! Luther was a great grandson of Jacob Pritts.

On February 8, 1907, the following event occurred and was published in Book 2 of The Rockwood Area Historical Society.

Law Man Rescued[9]
FEB. 8, 1907

Mr. Noah Pritts, of Wilsoncreek, was a jolly visitor at our office Tuesday. Noah recently sold his Somerset property to George Sechler and consequently feels prosperous. He told us about a thrilling experience be had rescuing Constable Ephraim Romesberg from drowning at Hay's bridge the other Saturday night, when the waters of Cox's creek were overflowing the Hay meadows and the public road. He was driving homeward in his buggy and Mr. Romesberg was following in his own rig. Just East of Rockwood a freight train blocked the Somerset and Cambria branch and Mr. Romesberg persuaded him to go around by Hay's bridge. He knew Eph's best girl lived on the upper road and he drove ahead just to give the young man company.

To get to the Hay's bridge Noah's horse was obliged to almost swim, but he got through safely and landed on the bridge. Not so with Eph, however. His buggy got mixed up with a raft of floating logs and his horse tore its harness in an effort to extricate the buggy from the drift wood. Ephraim called for help and then Noah's experience began. He tied his horse on the bridge and went back to rescue Ephraim. Noah never hankered for ice water in mid-winter, but on this occasion he waded in the flood up to his arm pits to get to where Eph was. It was bright "moonlight," which helped much. He disentangled Eph's conveyance from the dangerous flotsam of logs and saplings and made the rescue, for which he has the young constable's everlasting gratitude, particularly since Ephraim did not get wet in the least; and told Noah the best thing in the world to have is good friends. *Taken from a newspaper article.*

Chapter 4

Solomon and Polly Move to the Farm

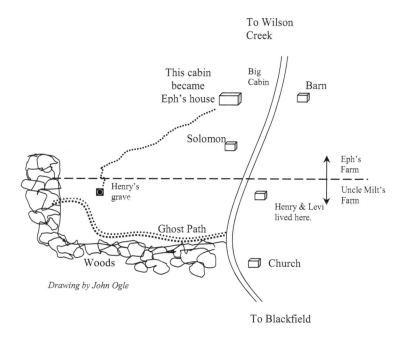

Drawing by John Ogle

At the time of Henry's death, Solomon and Polly, along with six children, lived in a small one-room home on Route 653, which runs from Rockwood to Garrett.

Henry had lived in a cabin on the property where his friends buried him. At the time of Henry's death, Jacob Pritts likely lived in the large cabin which most likely he built. Henry's sister, Polly, became the new owner of the farm. Polly later transferred ownership to her son, Levi. Levi married when he was 23.

Initially Levi and Mary Ann moved into the small cabin where Pop was born in 1878. This cabin was on the property when Henry was buried. Likely, Solomon and Polly moved into the second small cabin. Sometime later, Levi moved his large family into the large cabin that was converted into the farmhouse which still stands. Levi likely played a key roll in converting the cabin to the farmhouse.

The small one-room house on Route 653 where Solomon and Polly lived stood until some time after 1932. My sister Luella lived there for a short time after she married Woodrow Romesberg. I remember seeing the little house, which we called Solomon's Shack.

Levi and Mary Ann did well on the farm. Their total acreage was about four hundred fifty. Sometime, it is not clear when, the large cabin was converted into a frame house. Also, it is not known who remodeled it or who built the original cabin. The cabin was about 30' by 20', and the walls of this original cabin were made of studs measuring 4 inches by 14 inches by 16 feet. This space became the kitchen in the new house.

Solomon's move didn't create a housing problem. There were four girls and two younger boys, Levi and Samuel. The girls married and left. Levi soon had a large family (ten children lived), and needed the larger house.

Levi was born at a very opportune time in America—the railroad was born, coal was discovered, wagons were needed everywhere, sawmills were popping up everywhere, and farm crops were in increasing demand. Unemployment was nearly non-existent.

As time went by, the Pritts-Romesberg bond became stronger. Two marriages helped considerably.

- Polly Pritts' (daughter of Jacob Pritts) marriage to Solomon Romesberg, and
- Harrison Pritts' marriage to Barbara Ellen Romesberg (daughter of Levi and Mary Ann Romesberg).

From these two marriages, a lot of cousins were born in Black Township.

One of my favorite visitors to our farm was Eliza Pritts, Luther's mother. She helped my mother and Mom gave her milk and eggs. One day while I was working on baby chick houses, I looked across a small oats field close to the house. Eliza was fighting her way through blackberry branches to avoid walking through the oats field. I called to her and said, "Come back and follow the path." A group of neighbor boys had already trampled a path through the field. Eliza was very considerate of others.

Chapter 5

A Barrel of Flour

Sketch by Carol Ogle

I often think of the flour barrel on the back porch. I believe it was made of English tin. The old washing machine sat right beside it. One day while Mom was outside hanging up clothing, I decided to wash a pair of socks. I had a little trouble getting the socks in the wringer rolls. As a result, the rolls grabbed my fingers and pulled my right arm into the rolls. The rollers pulled me off the floor up into the tub of water. I hollered for Mom, as I watched my arm disappear between the rollers up to my elbow, and a fuse blew. Mom came to my rescue and opened the rollers. My arm looked a little like a pancake. Mom called my oldest sister, Luella, and asked her to go to town for new fuses. That evening Mom told Pop I had done a foolish thing, then reminded him to buy two bags of flour. Pop then said, "Grandpa Levi also did a foolish thing one time. He made a bet he could carry a barrel of flour from Garrett, 3½ miles, to the flour barrel on our porch without letting it down to the ground to rest. A barrel of flour weighs 196 pounds."

Levi had gone shopping with horse and buggy alone sometime in October, 1888. Grandma Mary Ann stayed home because Aunt

Della was only two weeks old. Levi had done well. He had every item checked except one—a barrel of flour. He had just enough money left for the flour. So he unhooked the strap from the tether pole and headed toward the mill. He was about there when he noticed a flask of whiskey in a store window. Just for curiosity's sake, he stopped to look at it. The glass was a faint green and filled with bubbles. The clear liquor inside looked cool and tempting, so he bought it.

As Levi left the store it suddenly dawned on him that he had no money left for the flour. Levi liked to talk, so he thought maybe he could talk the mill owner into charging it. The conversation at the mill took several turns, but the bottom line every time was, "no more charges, poor credit." Levi was practically on his knees.

What would he tell Mary Ann? The kids all wanted bread for supper. Then Levi said, "If I had that barrel of flour, I would carry it all the way home on my shoulder."

Suddenly, the attitude of the mill owner shifted. "You mean you think you can carry a barrel of flour 3½ miles!"

"Yes, Sir, I do," Levi replied.

The mill owner countered, "And without letting it down so you could rest?"

"Yes, Sir, I do," Levi replied again. The mill owner seemed amused. "And what do I get if you lose?"

Levi, who was a good carpenter, had an answer. "I'll finish that chicken house you started six months ago."

The mill owner asked, "When will you do this?" The mill owner knew he had a good bet. "A week from the next Saturday," it was decided.

Levi left in a hurry. He took the long way home so he could stop at Harrison Pritts' house to borrow a bag of flour from his daughter, Barbara Ellen. He got home just in time for Mary Ann to bake hot biscuits for supper.

Levi made his plans on the way home. First he would tell Mary Ann the whole story, so he did. Mary Ann asked Levi, "Where is the flask of whiskey?"

Levi answered, "Well, I just put it in my pocket; it should be in the buggy where I left my coat." Levi went out to the barn in the dark, lantern in hand. He brought the coat into the house. He pulled the flask from the pocket; the seal was unbroken. Levi had been so preoccupied that he forgot the flask of whiskey.

Mary Ann said, "Give it to me; you won't need it to learn how to carry the flour."

What man could carry 196 pounds 3½ miles? Levi knew he could because he had carried railroad ties that weighed 200 pounds—but he didn't know about carrying that heavy a barrel so far.

Mary Ann and Levi designed a sack with a narrow diameter in the center to fit over a shoulder. They filled it with 196 pounds of flour, which Levi got from the mill. Levi started carrying it every day. The final preparation steps called for help from sons Eph and Alex. Levi talked to the boys, "Don't make a whisper to anyone until after the contest!"

In the meantime, Levi asked the mill owner for a copy of the rules for the race. The mill owner was a shrewd businessman. He told everyone about the bet he had with Levi. So others started betting too. People stopped at Garrett—mostly to see what giant of a man could carry a barrel of flour 3½ miles without letting it down to the ground to rest. The mill owner came to realize that he had already won the bet through all the free advertising he received. His sales went up overnight.

Levi was strong and durable. He could work sixteen hours some days. He knew how to pace himself. Some, who knew his work habits, bet on Levi.

At 10 A.M. sharp, the marathon began. Eph, a piece of paper in his hand, led the way, followed by Levi with his burden. Next came Alex. There was a crowd of about eighty people. As Levi slowly and deliberately followed Eph, he suddenly felt alone and his mind wandered. "How did I get myself into this ordeal? Have I kept my promise to my father to help pave a way for others to come to America?"

As usual, he wondered who helped Stephen. What happened to his parents? Did he drink too much whiskey? Suddenly he felt the need to rest. Eph looked back at his father. Levi winked at Eph as he turned off the roadway and walked directly toward a young apple tree with branches about 4½ feet from the ground. As Levi stepped toward the tree someone in the crowd yelled, "Look, he's leaving the roadway, he can't go any further, he loses his bet!"

"Not so!" replied Eph, as Levi shifted the load from his shoulder to the limb. "Read the agreed upon conditions of the bet," Alex said to Eph. Eph pulled the paper from his shirt pocket.

"He must not lower the load to the ground to rest." Everyone started milling around and muttering. Some made new wagers. Finally, Eph pulled a second piece of paper from his pocket and called out that the next tree was only five minutes away. Levi seemed to get stronger the further they marched. They stopped at every tree marked on the map Eph carried.

Finally they came to the little church. Everyone knew he was soon there. Now they could hear shouting up ahead. A large crowd of people crowded the yard in front of the log cabin. As Levi approached, the crowd opened up just enough space for a straight path for Levi to empty his barrel of flour into our barrel on the porch!

A week later, as the mill owner left his house, he heard the sounds of a hammer and saw. "What are you doing Levi?" the mill owner called out.

"You still need that chicken house, don't you?" answered Levi.

Levi went about his daily work and helped plan his second daughter's wedding. Emma was married on March 19, 1889. Levi stayed up late after the wedding. The next day, Eph got up early and came down to the kitchen. He found his father lying with outstretched hands over the kitchen table. He had an empty flask in his hand. He had died of a heart attack.

Chapter 6

The Spoke Factory and the Grey Squirrels

Sketch by Carol Ogle

Sketch by Carol Ogle

Levi's untimely death on March 20, 1889, left his family in dire straits. The biggest problem was debt and lack of adequate income for a large family.

Levi and Mary Ann's Family	Age at Levi's Death
Barbara Ellen 1865–1902	24, married
Joel 1867–1938	22
Emma G. 1869–1908	20, married
Alexander 1871–1958	18
Hiram (Davy) 1873–1924	16
Levi, Jr. 1875–1957	14
Nettie 1877–1877	–
Ephraim J. 1878–1964	11
Milton Warren 1881–1963	8
Joseph 1883–1968	6
Ralph 1885–1969	4
Della May 1888–1984	6 mo.
Mary Ann 1844–1929	45
Levi 1840–1889	49

There were eight children at home when Levi died. It was a time for Mary Ann to step forward as head of the household. Eph said it was in November of that first year when he went to the basement and found they had only a bushel of potatoes. He rushed upstairs and called out to his mother, "Mom, we're going to starve! We have only one bushel of potatoes."

Mary Ann replied, "No, we won't starve."

"How do you know?" Eph asked.

"I know because I am praying and reading the Bible every day," was her answer.

Mary Ann called the boys together and they laid out plans to save Levi's farm:

1. Joel and Alex would take out a loan for as much as the bank would give them.

2. Mary Ann would work with the bank to assume responsibility for part of the debt.
3. They would sell some land to pay off the remainder of the debt.
4. Mary Ann would manage the farm.
5. The four older boys would get jobs."

Ephraim dropped out of school when he was eleven years old. He got a job in the spoke factory, which was about a mile away in the woods. He received 10¢ a day. He cut white oak for spokes for wagon wheels and handles for tools. The boys decided to hunt squirrels for the table. Mary Ann approved. They shared in the cost for the shells. They had only one gun so they shared its use. They shot squirrels through the fall and most of the winter. Ephraim many years later told us it was a miracle. It so happened that in spring there was a late frost that froze the blossoms on the oak trees. However, the frost was spotty and did not kill the blossoms on the trees on their farm. So squirrels migrated to the areas where the acorns were plentiful. They usually brought home squirrel for supper every day. They never went hungry. Mary Ann's strategy worked out quite well except for one consequence: the size of Levi's farm was reduced from 450 acres to 250 acres.

When I left for school, I said that there is one thing I would like to have from the farm—a picture. It now hangs on my dining room wall. On the back of the picture in big letters, my wife, Shirley, wrote, "FLOYD'S GRANDMOTHER."

I've often had these thoughts:

My Grandmother Mary Ann's picture hangs on my dining room wall. She stands there in front of her garden picket fence and flowering apple tree. She stands there every day with prayer looking out for me.

*Mary Ann Livingston Romesberg. Born April 2, 1844, married to Levi
Romesberg, born October 31, 1840. Parents of Ephraim Romesberg who
married Mayme Susan (Swearman) Romesberg.
Photograph from family records.*

Black Twp Echoes, July 4, 1976, Page 9

Musicians Entertain Romesberg Clan[10]

SEPT. 22, 1916

Surrounded by her children, grandchildren and great grandchildren and neighbors and friends, Mrs. Mary Romesberg was perhaps the happiest and proudest resident in all of Black township last Sunday.

The occasion was a family reunion of the Romesberg clan, of whom Mrs. Romesberg represented alone a generation, and three additional generations figured conspicuously at the reunion. The scene of the celebration and its accompanying festivities was the old Romesberg homestead, where Mrs. Romesberg has resided throughout nearly all of her married life and where since the death of her husband, the late Levi Romesberg, she has continued to reside with her son and daughter-in-law, Mr. and Mrs. Ephraim Romesberg. This estimable woman, now in the seventy-third year of her unusually active and useful life, is, as some one aptly remarked at the reunion "as spry and cheerful as a girl still in her teens." Notwithstanding that the younger of her kin insisted that she sit idly in the shade and enjoy the big function that had been staged in her honor, the venerable lady was among the busiest, planning and working for the comfort and pleasure of the numerous guests present.

The Wilsoncreek Cornet Band, one of the best musical organizations in Somerset County, was on hand and, under the able leadership of James J. Miller, played a number of concerts during the day.

A big feast, in which many of the excellent cooks of the Wilsoncreek community took a hand, was served on the lawn at the Romesberg residence and "In the Shade of the Old Apple Tree." The table reached all the way across the lawn and even a chronic dyspeptic would have capitulated before the display of appetizing edibles. Just as the feast was fairly under way a foot-sore and ravenously hungry tramp came along, probably having scented the savory victuals from afar. Naturally, Grandmother Romesberg invited him into the yard and with amusement and delight watched the half-starved wayfarer "pitch in," which he did without awaiting the second invitation.

After addresses by Messrs U.W. Werner and W.H. Coughenour, the latter photographed the relatives according to generations represented and also as an indiscriminate group.

Of Mr. Romesberg's family there survive six sons and one daughter. There are also 39 grandchildren and 32 great grandchildren.

The Miner's Band Wilsoncreek[11]
Circa 1915–1920

Left to right, back row: Charles "Espy" Weimer, Perry "Aust" Weimer, and James "Jim" Pritts. Standing in the second row: George "K'nail" Pritts, Joseph "Joe" Romesberg, Edward "Ed" Atchison, Ephraim "Shorty" Weimer, Charles "Cappy" Pritts, and Ralph "Feltie" Romesberg. Seated in the third row: Ephraim "Eph" Romesberg, George "Sticker" Playze, Joseph "Joe" Pritts, Samuel "Rich" Weimer, Edward "Ed" Baker, and Calvin "Cal" Atchison. Seated on the ground is James "Jimmy" Miller.

A Brass Ensemble[11]

A brass ensemble consisting of Ephraim Romesberg on coronet,
Ephraim "Shorty" Weimer and Alvin Atchison (both seated) on baritones.

Chapter 7

Wilsoncreek and Blackfield

Wilson Creek Miner's Band, circa 1900[12]

This photograph was taken in Rockwood's lower diamond.
Drivers seat: Harrison Pritts (standing, holding reins), Alex Romesberg;
left to right: Cal Atchinson, Geo Pritts, Jimmy Miller, Ralph Romesberg,
Chas. Pritts, George (Sticker) Plays, Ed Atchinson, Joe Romesberg,
Joe Pritts, Chas Weimer, Aust Weimer, Shorty Weimer, Rich Weimer,
Ed Baker. Photograph courtesy of the Rockwood Historical Society.

More than a hundred years before the present date (1883), James Wilson executed a cabin on a tract of land about 2½ miles east of the present town of Rockwood.[13] Quite probably he was the first permanent settler in the township. Wilson had a sawmill on his farm, which was the first manufactory in the township. Other early settlers were: Peter Bowlin, George Enos, U. N. Nicholson, Jacob Snyder, Jacob Weimer, William Critchfield, John Dull, Frederick Weimer, Francis Phillippi, and others. The first township election was held at the home of John Shaff.

James Wilson built his cabin close to the small stream. Later, Uncle Joel's barn and house were built at this same location. The

barn still stands and is located between Route 65
The railroad spur (from the main line which runs
Summerset) and the creek cross the road at the sp
stood.

Barn on the Romesberg Farm[12]

*The farm is located between Route 653 and Mud Pike. On horseback are
Joel and Harvey Romesberg. The farm is now owned by Mary Ann and
Ted Fundis. Mary Ann is the daughter of Harvey, and Harry is her brother.
The barn still stands.*

John Dull, a hunter and trapper, was one of the earliest pioneers in
our area.[13] He was born in 1753, and came from Eastern Pennsylvania
when a young man. He took up a considerable tract of land in the
northeastern portion of Milford. In the course of time, Dull divided
his land into four farms, one of which he gave away in order to get
a neighbor. He was obliged to go to the vicinity of Bedford for milling, going and returning with packhorses through the woods. In his
hunting exploits, he would often camp on the mountains, and remain
there alone for weeks. On one occasion he and his dog Tiger ("Tige"
for short) treed a mountain lion. Dull fired and wounded the beast,
which immediately sprang from the tree and attacked him. As was
his custom, he carried a butcher-knife and a hatchet. He grasped the
hatchet and struck at the mountain lion, and the hatchet came off the
handle, leaving the latter in his hand. The savage beast rushed furi-
ously upon Mr. Dull, and tore his hunting-shirt from his body. Tige
came bravely to his master's aid, and attacked the mountain lion in

ar. This movement diverted the mountain lion's attention, and ull found an opportunity to use his knife with good effect. After a long and terrible contest, the mountain lion was killed. Mr. Dull was terribly scratched and bitten, and Tige was so nearly used up that he had to be carried home.

Wilsoncreek and Blackfield grew as the coal business developed. There were about 60 houses built in Blackfield and 50 in Wilsoncreek. At one time, about 400 people lived in Wilsoncreek. The railroad spur off the Rockwood -Somerset line follows Wilsoncreek all the way past No.4 and No.10 mines in Wilson Creek, to the tipples in Blackfield.

The now deserted little town of Wilsoncreek and the small stream was named after James Wilson, the first resident of the area.

When my father, Ephraim, was still quite young, he dreamed big monsters appeared along Wilson Creek. Soon thereafter, over a distance of about a mile, five coal mines were opened.

Each mine had its own tipple. A tipple is like a big funnel used to dump the coal into the railway cars. A string of cars were inspected for rock and slate and dumped at the tipple into the railway cars. Slate and rock are dumped out onto the slate and rock pile. A second dumping station was used to dump coal into horse-drawn wagons for home use or into trucks for delivery. All homes were heated with wood and coal.

A lot of timber was used in the mining operation. Pop cut trees of about 5" to 6" diameter and some 6" to 7" in diameter. Dinkey tracks ran as much as one to two miles underground. Mining posts were used to hold the roof from falling in on the miner. Wilsoncreek usually had 7' cross beams and 32" posts. Pop cut most of the timber for Wilsoncreek over a period of forty-six years. However, I got the last order for the Wilsoncreek mine in 1946, which paid for my college expenses.

Blackfield processed their coal in a uniquely different way. Cars loaded with coal were pulled out of the mines with mules and then pulled on a dinkey track about 1½ miles to a tipple located on a railroad spur along the Casselman River in a deep valley with the Negro Mountain to the south and the River Hill to the North. Two

**This is Wilsoncreek before the turn of the century—
before a road was put in place.**[9]

*Consolidation Coal Company – Mine No. 4 & Mine No. 10
Photograph courtesy of Rockwood Historical Society*

**Wilson Creek
Slate & Rock
Pile**[9]

*Charles Neimiller
(center) with two
unidentified help-
ers at the top of
rail car—around
1900 (what a job
they had!)*

Photograph courtesy of Rockwood Historical Society

dinkey tracks ran down the River Hill. The two cars, one car loaded at the top and one empty at the bottom, were hooked together. Using a brake on the cable, the loaded car is run down and its force pulled the empty car back up.

Wilsoncreek had a unique water supply for the houses. Water was collected in a reservoir and piped down-hill to the houses through staved wooden pipes held together with metal clamps. This pipe is still solid after carrying water for 100 years. Merle Romesberg II took the following picture. The photo shows the open reservoir with water still running into the wooden pipe.

Wilsoncreek Reservoir – July 2007

Photograph by Merle Romesberg II
Wilsoncreek Water Reservoir (above) is located on Fogles Coal Company land above Wilsoncreek Mine slate pile. You can see the place they dug a ditch to place wooden pipe going towards Wilson Creek. You can follow the ditch through the woods down to the slate pile. Wilsoncreek Water Reservoir is 39 Degrees 55.13N minutes 079 Degrees 05.54W 2152 Ft minutes using NAD27 Datum. (39 deg 55.13N, 079 deg 05.54W, 2152Ft)

Wilsoncreek had a two-room schoolhouse. By the time we went to school, only one room was used and eight grades were taught.

My best friends in school were Norman Hay and Bill Shaw. The teacher had a hickory stick on two nails along one black board. We thought it was just for show. Bill got into trouble one time. The teacher lost control of himself and grabbed the hickory stick. He took Bill by the arm and pulled him out to the second room. Bill screamed and hollered in pain. That evening, a car pulled up to our house. Out of the car jumped Bill's grandmother, Mrs. Keeler. "Where's your Pa?" she said, looking straight at me.

"In the house!" I said.

"Go tell him I need to see him."

I didn't waste any time. Pop came out to the car and Mrs. Keeler reached in the back seat and pulled Bill out of the car. Then, she lifted his shirt.

"Wow!" I said to myself. There were big, long, and bloody welts all over his back. Pop talked to Mrs. Keeler, then she drove away.

The next day at school, I sat looking out the window. Suddenly, I saw someone coming down our path. Just as the path opened up into the rocky playground, I saw it was Pop. I thought maybe he and the teacher would get into a fight. There was a light knock on the door. Our teacher was with a class. He got up and walked to the door. Everyone was quiet. Then our teacher came quietly back into the room. He walked up to the blackboard and reached up and pulled down the hickory stick. He took it to the door them came back, empty handed.

I stared out the window. Pop walked across the playground and just as he got to the tree line, he stopped, turned, and looked right at the school. Then he raised his hands and broke the stick into two pieces and tossed them into the woods! That was the last we saw of the hickory stick.

Bill moved to Johnstown but remained a good friend. For a while, he cut a comic strip of "Tom Sawyer" out of the daily paper and mailed these to me once a week.

I was in fourth grade when we started a softball league, and I played ball with my brothers about every Sunday afternoon. I pitched for Wilsoncreek every game but one, for five years. Butzz (Earl) Pritts pitched one game. We lost only one game to Blackfield in those years. I pitched that game.

Much of Blackfield and Wilsoncreek is history. Houses were built and houses were torn down. Today, stone walls are all there is left of the houses and the school. Many of the stories of old are lost because they were not recorded. Some few have been recorded. Others are still in the minds of those who heard the story last.

Another mountain lion story was about to start in 1937. Paul Pritts saw one pass in front of his car at night near Matt Romesberg's residence. Later my cousins Lynn and Tom heard one scream near our farm. It so happens I had gone fishing one evening. There was no moon and it got very dark. I quit fishing and walked to the road and just then I heard voices coming up the road. They came very close to me, but couldn't see me in the dark. I thought about the mountain lion and just as they passed me about five feet away, I let out an improvised mountain lion scream. Everything was dead quiet. The boys stopped at Matt's and were afraid to go home alone.

A couple of weeks later, the boys' Dad, Sam, stopped at Wilbur's for a gallon of maple syrup. Sam told Wilbur about a mountain lion running loose. Wilbur then told Sam, "That wasn't a mountain lion, that was Floyd." Wilbur said Sam couldn't stop laughing.

Uncle Alex played an important role in our family. Uncle Alex and my dad both played in the Wilsoncreek Band. The band members rode in a horse-drawn wagon. Alex was engaged to be married when his fiancé died. He never married. When Levi died, Alex assumed the loan on the Long Farm, which he later sold to Zimmerman. He lived with Mom and Pop.

Alex liked to hum while rocking on a rocking chair. When he hummed, he closed his hands over his head. One time when Alex was humming with his back toward an open window, Merle reached in the open window and lifted his hand in the air with a broom stick. Uncle Alex rushed out of the house and gave Wilbur a licking. He didn't know which boy was the prankster.

When the house got too crowded, Alex bought a house in Wilson-creek. Alex continued to come to our house to work and to eat.

He always turned the meat grinder to make sausage. He seldom missed helping to make hay.

Alex bought the Bittner farm for Eph's five boys. This farm became Merle's dairy farm after WWII.

Uncle Milt worked in the Wilsoncreek Mine until he had an accident. His one leg became shorter than the other, and he used crutches the rest of his life.

After his accident, Milt opened the general store in Wilsoncreek. We traded eggs for groceries at the store. Milt always drove a little truck. He gave us rides often. This ride made carrying groceries a lot easier. Milt bought property from Pop. His house was built across the road from the log cabin where Levi's family had lived. Henry Pritts was buried on Milt's farm.

When I was writing about Uncle Milt, I called my first cousin Minnie, uncle Milt's daughter. She then sent me a letter with a copy of a short autobiography, which Uncle Milt wrote, not long before he died.

Letter from Minnie Romesberg Thomas

Dear Floyd,

The following brief autobiography was written by my father while he was confined to bed following a stroke which he suffered in February 1954. Later he gave the copy to me. It was written on ordinary lined tablet paper in pencil—probably sometime in the year 1956. This typed copy is just as he wrote it, except that I supplied capital letters and some punctuation marks where necessary to make it more legible. As you will learn when you read it, my father had very little formal education, so he was not very knowledgeable in the fundamentals of English grammar. He had told me that he was unable to speak English until he entered school, having spoken only Pennsylvania Dutch up until then. He was largely self-taught and later became skilled and knowledgeable in many ways.

I put the written copy that he gave me in my desk when I returned home, and had forgotten about it for many years. Then in the spring of 1993 when I was cleaning out my house in Philadelphia and preparing to move into the retirement home where I am now living, I came across it. The pages had become brown with age, and I was afraid the paper would become brittle and deteriorate. I thought then that I should do something to preserve it as part of my heritage. But again, after I moved and finally settled in my new home, I forgot about it until someone asked me for some information about my parents. Then I began to search for it, and for some time I was unable to remember where I had put it. Then the thought occurred to me that I may have placed it in my copy of "The Romesberg Family History," which was a logical place for it—and there it was. I decided then and there to have it typed and to send a copy of it to each of my nearest relatives.

I hope that you will appreciate it and will preserve it as part of your history and heritage. And I hope that it will be passed on to future generations.

Some of you probably have a copy of "The Romesberg Family History," which traces the genealogy of the Romesberg family. This research was largely the work of my first cousin, Blanche Snyder, the daughter of Scott and Della Romesberg Snyder. Della was my father's youngest sister. Blanche devoted many years of her life to this research, and we are indebted to her. Her information was compiled by a special committee and was published in book form in 1984. If you have a copy, I suggest that you keep this document in the book as an additional part of your family history. Otherwise keep it in a file or some other safe location.

(Incidentally, in "The Romesberg Family History," there are a couple of errors in the information listed for me and my family under the number (521). My birth date is listed as February 20, 1914, but the correct date is February 26, 1914. Also, my son, John Milton, was born on July 14, 1949 (not 1940 as listed). Please correct these errors in the book if you possess a copy.)

My father died on August 4, 1963 at age 82, about 9½ years after his crippling stroke. My mother nursed and cared for him all those years, in spite of severe arthritis. She died on August 28, 1970, at age 86.

Sincere good wishes and may you be blessed with good health and good fortune.

Minnie Romesberg Thomas

Milton Warren Romesberg, son of Levi and Mary Ann Livingston Romesberg, was born March 14, 1881, in an old log house in Black Township, Somerset County, and lived in the same Township all his life. His father died when Milton was only eight years old, leaving his mother a widow with a large family and a large debt. At the age of thirteen, he was required to quit school and go to work in the woods to help support the family, working in the timber until the age of 19. Then he changed work and got a job in a coal mine at Wilsoncreek and worked there five years until the age of 24, when he was compelled to quit his job and be taken to a hospital in Johnstown. While working in the coal mines, he contracted what was then called hip joint disease of the bone. After a long illness this disease was finally checked, but Milton lost the use of his hip joint entirely and walked with a crutch, and often two crutches, for twenty-five years. On account of this disability he could not work in the mines or at any strenuous labor. In 1931, and again in 1932, he was taken to the Somerset Hospital for an operation on his hip, which had become infected, evidently because of strenuous work. In 1951 he was taken to the same hospital for the treatment of a severe case of rheumatism, a broken shoulder, and pneumonia. But worst of all, on February 19, 1954, he was stricken with a cerebral hemorrhage and was taken again to the Somerset Hospital in a serious condition, remaining there fifty-two days with the care of two doctors and three trained nurses. Now, two years and five months later, he is still not able to walk except with a walking chair, and is still suffering much pain in his head, face and the right side of his body.

Autobiography by Milton Warren Romesberg
(Provided by Minnie Romesberg Thomas)

I was married on June 11, 1901, to Miss Emma May Weimer, daughter of Cyrus and Trucilla Baker Weimer, by the Rev. John T. Balliet, a pastor of the Reformed Church in Rockwood, at the home of Mr. and. Mrs. Perry Baker in Rockwood, PA. We started housekeeping in the early fall of 1901, renting three rooms in the Noah Pritts' old family home, then owned by my brothers Joel and Alex. It was located near the pit mouth of the Wilsoncreek Mines, where I was employed. Here our first baby was born. Early in the spring of 1902 we moved in with my brother-in-law, Harrison Pritts, whose wife (my oldest sister) had died in March of that winter. Here we lived until our new home was finished in October. Then we moved into this, our own home, "built squarely across the road" from where the old log cabin stood and where I was born. We have lived here in this home ever since.

We are the parents of five children: four boys and one girl. Earl Clinton was born November 8th, 1901. Howard Graydon was born June 15, 1903. Gerald Alvin was born April 9, 1909. Roy Vernon was born January 13, 1912. Minnie Melda was born February 26, 1914. Earl is married to Merium Lingg of Hanover, PA. He is a medical doctor, has practiced medicine since 1930, lives in York, PA. They arc the parents of two children, one boy and one girl.

Barbara, the oldest, is married and is the mother of two sons. She is married to a young medical doctor and lives in the State of Washington. Her husband is serving his Country in an Army hospital there. Earl Jr. is married and is the father of two sons. He served three years in the U. S. Marines and now lives in Cleveland, Ohio, and is employed by Republic Steel. Howard is married to Pearl Miller of Black Township. They are the parents of three daughters and live in Black Township. He is employed as the janitor of the Rockwood High School. Their oldest daughter, Bernice, is married and lives in Warren, Ohio. She is the mother of one son and one daughter. The other two girls are still at home. Gerald is married to Edna Bittner of Summit Township. They are the parents of five children, two boys and three girls. They live in Brothersvalley Township on a dairy farm. Their two oldest daughters, Patsy and Doris, are married and they both live in Akron, Ohio. Doris is the mother of one baby girl. Roy Vernon died on March 15, 1912, at the age of two months. Minnie was a high school teacher for twelve years and is now married to Basil Thomas. They are the parents of one boy. They live in Philadelphia, PA.

On November 1, 1906, I was discharged from the Johnstown Memorial Hospital, and after recuperating for one year at home, but still walking on crutches, I hired myself to Noah Pritts to work in a small grocery store at Wilsoncreek, my salary being $12 per month and board. I worked for Mr. Pritts for ten months and then purchased the groceries from Mr. Pritts, paying $420 for same, which amount I borrowed from my brother Alex, giving him my note for same, which I paid in small payments years' later. One year later I purchased the store building from Mr. Noah Pritts. The purchase price was $225 cash. Sometime later I purchased the lot, on which the building was erected, from William Ephraim and Fannie Pritts for $30 cash. Then in 1914 I built a larger store room building on an adjoining lot which I purchased from George Pritts. Cost of building was approximately $1100. I do not remember the price paid for the lot. Here I was in business for myself until I was stricken with a stroke on February 19, 1954. Nearly forty years in the larger building and seven years in the first building, making a total of 47 years in store business. I also served my church and community for many years, serving my district as tax collector for twenty years, as Judge on the Election Board and Township Auditor for many years, a charter member of the Rockwood Union National Bank Board serving 21 years, also Elder, Deacon, Sunday School Superintendent, and teacher of a Sunday School class in St. John's Lutheran Church and Sunday School for many years.

On June 11, 1956, we celebrated our 55th wedding anniversary.

Milt sold dynamite, which was a big product for income. The biggest use for the explosives was in coal mining. Milt had a powderhouse on the back of the property. The building was made of solid concrete. It had a big crack on one side. We were afraid to get close to the powderhouse. The powder house for the Blackfield mine exploded. See following newspaper article, copied from the *Berlin Record,* published July 19, 1918.

Blackfield
Wrecks Town and Kills Man
U. S. BUREAU OF MINES ORDERS INVESTIGATION
Man's Body Blown in Many Directions—
Believed Watchman's Carelessness Caused Explosion

The powder magazine of the Atlantic Coal Company at Blackfield, in the Rockwood region blew up Wednesday morning between 1 and 2 o'clock, killing one man, wrecking part of the village and more or less damaging every building in that place. The explosion was heard more than five miles away and the light was seen for a great distance, some of the men on night duty at MacDonaldton claiming to have heard the noise and seen the light. Where the magazine stood there is nothing left but a hole in the ground; the engine house, a couple of hundred feet distant, was smashed into a pile of rubbish; the glass fronts of the store and hotel, 300 feet away, were blown in and the goods in the store were piled from the shelves onto the floor. The sides of the other buildings were blown in and the rafters of still others were broken. The windows of about all the buildings in the place were shattered to pieces. There was one noticeable freak, the electric station of the high power tension line escaped serious damage, though it was the nearest of any buildings to the powder magazine. A woman and her newborn babe in a house that was badly wrecked escaped unhurt, other than shock.

One man, George Saluski, the night watchman, was killed, and it is supposed that through some mishap of his the explosion occurred. The time for filling the powder cars was 6 o'clock in the morning, so it is said, but it is thought that Saluski was in the powder house when the explosion occurred as his watch, when found, showed that it had stopped at 1:40. It is supposed that he had gone into the powder house at that time to fill the cans, and that he had used an open lamp or was smoking. The real cause of the explosion will probably never be known.

Saluski's body was found scattered about in different directions. The main part of the body was found wrapped about the dinkey engine at the engine house; part of the skull was found 300 feet off in another direction. The liver was driven and wrapped about a tree trunk so tightly that it was necessary to use a sharp instrument in cutting it loose; an arm was found in still a different direction, as were other parts of the body. The fragments were gathered up and taken in charge by a Rockwood undertaker. Saluski leaves a wife and seven children, one of them a babe that was born on Sunday. It is feared that the shock will kill his wife.

Norman Hay's dad died at ago 40. He died from black lung disease. Many miners got black lung from years of breathing coal dust from the everyday exposure to the dust.

Miners dug tunnels and rooms and removed the layers of coal. They laid metal tracks for the dinkey cars, which were loaded with the coal they dug from the seam of coal. Their light was the little carbide light hooked to their hats, which grew heavy as the day wore on. They had to stop often to add the calcium carbide in the lower compartment and water in the upper compartment.

Shocking, Deplorable Death[10]

A shocking, lamentable accident occurred at 7:25 p.m. Monday evening at the Romesberg stone quarry, near Atlantic station of the B. & O. R. R. and in Black Township, which almost instantly caused the death of Mr. Irvin Pritts, of Wilsoncreek, also in Black Township. Age 40 years.

Mr. Pritts and Mr. Edward Sheeler of Garrett, were working about twenty feet from the opening of a passage underneath a layer-like formation of rock. Mr. Pritts was in charge of a drill, which he held in desired position while Mr. Sheeler applied compressed air with which to operate the drill. Dynamite had been previously discharged near where they were engaged in drilling, loosening overhead, it is thought, a 1900-pound boulder, which without warning let go and fell upon Mr. Pritts crushing his skull and otherwise injuring him.

Mr. Sheeler himself had a narrow escape, since he told Coroner Emily K. Fluck at an inquest, which was conducted Monday evening in the office of Funeral Directors Mills & Mickey at Rockwood, that the falling rock "grazed my toes as it fell to the ground and crumbled."

Mr. Sheeler and other workers as rapidly as possible removed the more or less shattered rock from the body of Mr. Pritts. He lived only a short time after he had been extricated, and before a summoned physician could arrive.

The coroner viewed the greatly deplored death of Mr. Pritts as accidental, after having heard a number of witnesses and she stated that unless there are further developments, no formal inquest will be held.

Mr. Pritts was one of the most estimable citizens of Black Township. He was an industrious, useful, helpful neighbor and friend. In civic matters he was public-spirited, and in church having been a consistent member of the Sanner Lutheran congregation, whose edifice is in the vicinity of Wilsoncreek. His loss to that community is deemed irreparable.

Surviving Mr. Pritts are his wife and children, Evelyn, Charlotte, Marie and Albert, all at home. He is also survived by his father and stepmother, Mr. and Mrs. Harrison Pritts of Black Township.

He leaves four brothers, Messrs. George, Charles, James and Elmer Pritts, all of Wilsoncreek, and three sisters, Mrs. Mary Gorsuch and Mrs. Ephriam Weimer, both of Wilsoncreek, and Mrs. Nellie Mullen of Acosta, Pa.

Interment was subsequently made in the Rockwood I.O.O.F. Cemetery.

Western Maryland Coal Tipple[12]
The partial structure can still be seen from the Laurel Highlands Rails to Trails, west of Rockwood.

**Lumbering in Black Township Around
the Turn of the Century**

At left above is Harvey Livengood with horses.[12]

*Harvey Livengood is in the center of the group
dressed in long apparel – others unknown.*[12]

Wilsoncreek School
burned down.[12]

Rhodes School No. 1
*Located on Lloyd Snyder
farm. There was a Rhodes
School No. II also in
Black Twp.*[12]

Humbert School[12]
*Located off Mud Pike on the
Gorman Kincaid Farm. School
has now fallen down.*

BLACK TOWNSHIP

When Milford Township was divided in 1886, that part east of Coxes Creek and east of the Casselman River, below Rockwood, was given the name Black, in honor of Judge Jeremiah S. Black. Judge Black was later Attorney General in Pennsylvania's only President's cabinet (James Buchanon). Timber and lumber products, coal mining, the railroad, and aggressive agriculture developed in the area when our ancestors came here to start a new life.

First of all, our forefathers believed in one God, passionately and fervently.

Secondly, they believed in this land—America, unequivocally, and they believed in themselves resolutely. They were adaptable, determined, optimistic, honest, hard-working—and never depending on someone else to do the job for them.

Atlantic Fuel Company at Blackfield[12]
Located off Route 653, turn left to yard, then right. Pictured are Bill Heining, ? Barber, Frank Romesberg (operator of the company store), and Bill Miller.

Homer Shaffer,
Standing in front of the store at Blackfield, with his groundhog.[12]

Boarding House in Blackfield,
shown are John Kelly and "Solly" Albright.[12]

Miners at Blackfield
Ike Boden, Bill Heining, and Bobby Rubright.[12]

From July 5, 1917 news.
The fourth of July celebration at Blackfield was one of the biggest events in the history of that bustling mining village. An interesting feature of the affair was the dedication of a park presented to the town by the Atlantic Coal Company and named by the people of the town "Floradale" in honor of Flora S. Black, wife of the president of the Atlantic Coal Company, Hon. Frank B. Black, a state highway commissioner. The park consists of about twenty acres of woodland on the outskirts of the mining village. Several acres of this have been cleared out and converted into a beautiful picnic grove.

Chapter 8

Then There Were Eleven

E.J. Romesberg – 1940
Photograph courtesy of E.J. Romesberg

The house where I was born was remodeled several times to accommodate our growing family. The original house was a log cabin, about 20 x 30 feet in size. Then it was made into a three-bedroom house with the original cabin used as the kitchen.

The second remodeling was done by Pop and his brother Joseph. They jacked up the house and made a larger basement and eventually constructed a six-bedroom house. This was completed in 1925, the same year my sister Della died at age 5.

When the house just had three bedrooms, the space was allocated as follows:

- Uncle Alex had one room,
- The maid and two school teachers who boarded with the family shared a room,
- Mom and Pop shared a room with six children, as follows—two children shared a bed with Mom and Pop, four children shared the other bed,
- Grandma Mary Ann had a room downstairs.

I can remember seeing my grandma before she died in 1929. I was quite small, but I have a distinct remembrance of my sister Betty rocking my cradle as my grandmother bent over to touch me. She wore a black dress.

Another strong memory I have was when I was three years old. My bed was in a sort of wide hallway leading to my older brother's room and also to the attic. My bed was on the floor, because I had bed-wetting problems, making it easier to keep my bedding laundered.

One night I awakened to loud voices coming from downstairs. I jumped out of bed and ran to the top of the stairs. My whole family was downstairs. There was crying and I sensed a feeling of much confusion. I was frightened and wanted to go downstairs to see what was happening, but my nightclothes were wet!

I heard Pop call out to Wilbur to go get Aunt Emma. Then to Merle to go to the Harrison Pritts home to phone the doctor. Then he told my sister Luella to get some hot water.

Later I put on my overalls and went down to the living room where Mom was lying on the couch. Dr. Saylor was sitting beside her. She reached out and touched me as I hovered near her. I knew she must be okay when she smiled at me. She must have known how scared I was.

Apparently she had fallen down fourteen steps, broken her collarbone and had multiple cuts and bruises on her face!

Two weeks later, on November 25, 1929, I was again awakened. It was early in the morning, the sun was up and everything was quiet. My sister Elaine, just a year and two months older than I, took me by the hand and said "Come, I want to show you something." There was Mom in bed holding a new baby! He had curly hair! I knew Mom was all right! I knew right then that this new baby would come to be my best friend. Pop was standing by the window. Mom called out to him, "If you want a son named after you, this is it, because we will be having no more children!" Then there were eleven! So Ephraim Jay had joined our family!

A list of family members when E. Jay
was born on November 25, 1930.

Name	Birth/Death Year	Ages (on 11/25/1930)
Ephraim	1878/1962	52
Mayme	1892/1965	38
Grandma Mary Ann	1844/1929	Deceased
Mary	1912/1986	18
Luella	1914/ –	16
Merle	1915/2004	15
Wilbur	1917/2000	13
Della	1919/1925	Deceased
Helen	1921/2000	9
Paul	1922/ –	8
Betty	1923/ –	7
Elaine	1925/2005	5
Floyd	1927/ –	3
E. Jay	1930/ –	0

One of our fall activities was picking nuts. There were many "Hog" hickory nut trees on our farm. The squirrels usually got to most of these nuts before we did. This was all right because these nuts are very hard to crack, so we only picked a few. We all liked the hazelnuts the best. The best stand of hazelnut trees was in Adam Snyder's pasture field. We walked to this field by way of the railroad track that passed the Keeler farm to gather these nuts.

One day E. Jay and I filled our old sugar sacks with hazel nuts and headed for home. E. Jay was about 7 at the time. E. Jay seemed very tired and we stopped to rest frequently. I carried both of our bags as I realized something was really wrong with E. Jay.

As soon as we got home, I called to Mom. "You need to take 'Junior' to the doctor. He couldn't carry a little bag of nuts without getting out of breath!" I can remember seeing the back of our old car as Mom and Merle left to take E. Jay to see Dr. Musser in Somerset. They came home with the news that E. Jay had asthma, and would be tested for allergies. One thing he was allergic to was "chicken feathers." From the age of seven until he was seventeen he had allergy shots once a month. At that age he joined the Navy and served for seven years. He was in a different environment and no longer required his allergy shots.

One of my best friends, besides E. Jay, was Charles (Sonny) Weimer. When E. Jay joined the Navy, he and Bobby Ringer (our half-sister Mary's son) went to see Sonny to convince him to join the Navy. Sonny said, "No way!" Later Sonny was drafted and sent to Korea. I was going to school at the University of Cincinnati. I believe it was 1951, when I opened a copy of the *Somerset Daily American* and read "Sonny Weimer Missing in Action." His body was later shipped home.

E. Jay and I spent a lot of time together until I left for college in 1945. Most of the time we spent working together on the farm (see Chapter 12). After E. Jay left for the Navy, I believe we only saw each other about five times during the next thirty years. However, during the past ten years we've had family reunions in Rockwood every fall and have kept in touch with each other.

Merle joined the Army (1941) and retuned in January 1946. That summer we had a family photograph taken. We did not get together again as a family for about fifteen years.

E. J. Romesberg Family in 1946

Photograph from family records
Front Row: Helen, Lois Elaine, Mary, Mayme Susan, Ephraim J.,
Luella, Betty
Second Row: Merle, Ephraim Jay, Paul, Floyd, Wilbur

Romesberg Barn, 1916

The farm of Ephraim and Mayme Swearman Romesberg *is located above Wilsoncreek along Colflesh Road, below Blackfield Church. It is presently owned by Homer Colflesh and family. (Photograph from family records)*

Ephraim and Mayme Swearman Romesberg in their later years.[9]

*Views of the farm as it
appeared in 1997* [9]

Chapter 9

Red and Black Boots with Love

Sketch by Carol Ogle

One April day in 1934, when I was seven, Mom said to me, "Here, I have something for you." I opened a paper sack and to my surprise found a pair of black boots with a bright red band around the top of them. Then I said "This isn't my birthday and it isn't Christmas." Mom said, "You need them, put them on." So I put them on, and they were a perfect fit. Then, I went out and walked in the ditch which had a little water running in it from the rain several days earlier. After a while, I went into the house and asked Mom if I could go to the field above Uncle Milt's house where Merle was plowing, so I could show him my new boots. She said "yes." So I hurried past Milt's house, past his barn and field to the fence line of his property. There I crawled through the fence and walked straight toward Merle and the horses, "Nellie and Frank." "Look at my new boots, Merle," I called to him. Merle said they were perfect for rainy weather.

Then Merle said, "Come with me, I need your help." I followed him. The horses were glad to get a rest. We came to a big rock. Merle had dug ground away from the top and sides of the rock. He had a long, dry chestnut pole with one end under the rock. While he raised the other end, he instructed me to slide a small flat rock under the pole as close to the rock as possible. Then he said, "Now come and

51

help me push down on the pole to lift one end of the rock." While he held down on the pole, I shoved some small stones under the edge of the raised rock. We repeated this about six times, until Merle could slide a log chain around the rock. Then we unhooked the doubletree from the plow. Merle then gave me the reins and told me to bring the horses to the big stone. He carried the doubletree to the rock and then hooked the log chain to the doubletree and said, "Now drag the rock back into the woods when I say go." He pried down on the pole and said, "Go." I tightened the reins and called, "Get up, get up, Nellie, get up, Frank." Nellie pulled first, then Frank pulled, while Merle gave a quick push down on the pole. The rock rolled out of the hole. Merle said, "Keep going and steer the horses to the woods."

"Get up, get up!" I yelled. After the rock was out of the field, I called out, "Whoa! Whoa!" The horses were glad to stop. Then Merle held on to the log chain and I drove the horses back to the plow. Then Merle told me to put the reins around my neck and hold the handles of the plow and to keep plowing. Nellie walked in the plowed furrow ahead and Frank followed along her right side. I plowed until I reached the fence at the side of the field.

When we got home that evening, Mom said, "Where did you get all the ground on your overalls?"

Merle answered, "He got the ground all over himself helping me dig out a big rock which I could not do myself." He also told Mom that I drove the horses to drag the rock into the woods and to plow some. He made me feel proud. He told me it is important to "use my head" to "solve problems" or to "build a better mouse trap."

One day several years later, I decided to do just that—to build a better mouse trap. To make it better, I would make it look like a hole in the wall and work like a cat's mouth. Briefly, I cut two pieces of sheet metal that when bolted together looked like a pair of scissors with the blades turned to close like a pair of pliers. A half-inch hole was cut in the top piece. A short tension spring mounted on the handle side served to snap the scissors shut. A little stick about the length of three toothpicks, held the mouth of the trap open. The stick had a notch cut in it to fit against the inside of the half-inch hole. A piece of meat was tied onto the stick. When the mouse pulled at the meat, the stick came off and the jaws snapped shut. I made it look like a cat's mouth. It worked. In fact, it was a great "mouth" trap.

One day when I was looking for scrap iron to sell, I found some staves from an old wooden barrel. They should be good for something, I thought. Two weeks later, I made a set of skis out of two of them. I tried skiing that winter on a field behind Milt's Powderhouse. Then I sold them to my friend Bill Shaw for 50¢. It was a good deal for me.

The pair of boots that Mom gave me was one of the best presents ever. But I think that her singing was the very best gift, for all of us. She determined the time to boil an egg by singing the first stanza of "Abide with Me." She sang almost all the time. My favorite was a song sung to the tune of "Red River Valley."

There are songs that are sung by the prairie,
There are sweet songs and lovely melodies,
But the songs I will always remember are the
Songs that my Mother sang to me.
May I sleep in your barn tonight mister?
It is cold lying out on the ground.
The cold north wind is whistling and
I have no place to lie down.

I think of my mom every day. The following story is about my Mother and my favorite Christmas. Also included is a poem that Merle wrote when he was in the Army.

MY GOOD CHRISTMAS

The wind blew gently as millions of big snowflakes seemed to be racing to get to the ground first. My cheeks and nose felt gently cold as flakes hit me and seemed to tell me I was in the way, keeping them from hitting the ground and building up a layer for the big logging sled which the horses pulled to bring the big logs to the sawmill. Maybe Santa needed the snow on the ground, too, so his reindeer could pull his sleigh easier. I kept walking and making tracks in the fresh snow as I approached Wilson Creek, where there were more spruce trees. My eyes searched back and forth as each little tree was beginning to bend with the load of new fresh snow building up on the tender branches. Finally, I saw one I liked. I walked around it several times. I finally shook off the snow, then walked around it again. Then I cut it down with the little saw. I started back home, following my tracks that were already disappearing with fresh snow. I knew the way to our big white farmhouse, and I wondered if Mom would approve of my choice. This was the first year I went for the tree myself. I kept counting the days until Christmas—five more and then it would be Christmas morning. This was Saturday, December 20, 1935. I was eight years old.

This Saturday afternoon I helped Mom with the usual chores. I caught three roosters and cut off their heads, and Mom dunked them in boiling water so we could pluck off the feathers. Tomorrow would be our usual big Sunday chicken dinner with mashed potatoes and gravy. My sister Luella and her husband, Wood, came with beautiful little Peggy, who was less than two years old.

Then came Monday—three more days until Christmas. It stopped snowing on Sunday. The snow was about a foot deep except in the road. It was drifted shut in some places three to four feet deep. This was the day Mom and Pop went to Somerset with Matt and Lil to buy presents. Mom and Pop walked the mile out to the highway through the deep snow, and there the four of them crowded into Matt's little pickup truck. They got back well after dusk. I was already fast asleep. When I got up the next morning and came to the kitchen, Mom was sitting at the table drinking a cup of coffee. The oatmeal was ready for me. She said, "Eat a big bowl, it's cold outside." Then she said she and Pop came home well after dark and each had a big bundle of presents and she was afraid she dropped someone's present as they plodded through the deep snow in the darkness. So she asked me to follow their tracks all the way back to the highway and to look carefully to make sure someone's present wasn't dropped, and then one of the ten kids wouldn't have a present. So I did. Right from the porch door I started. I stepped in Mom and Pop's footprints in the snow, making sure I hit every one, and I looked carefully to both sides to see if a gift was dropped. I went past the summer house, the chicken house, through the oats field to the woods, past the pine trees to the road, then a field, an orchard, Harrison Pritts' house, then his barn, the big woods and the ball diamond, the Long farm, the spring along the road, and finally the highway where they got out of the truck. I stepped in all their footprints and looked to both sides—all the way there and back. I rushed into the kitchen, "Mom, you didn't drop a single gift."

Christmas morning came and we went to the tree early, the tree I had cut and dragged the long way home. There was my present. As I opened it, I thought of Mom and Pop's trip. They walked the one mile through the deep snow and then rode twelve miles in the cold truck to Somerset, then through the deep snow and darkness. They hung onto each child's gift, making sure none got dropped. My gift was a book—*Black Beauty*. I looked at the picture of the beautiful horse on the cover of the book. I knew I would treasure the gift. Not because of what it cost, but because of the love and care of Mom and Pop. How they walked through the darkness in the deep snow, each carrying a bundle of gifts and each making sure none was dropped so each of us ten kids would have a present Christmas morning.

My Mother

Mother has made me what I am today
To be kind and loyal and to obey
When I was young she taught me what was right
And took care of me when I was asleep at night
She got me ready and sent me to school
To learn and to apply the Golden Rule.
Though sometimes I may have caused her distress
Like a wandering boy in the wilderness
But today I am glad what she has done for me
Because it kept me out of trouble and misery
Now I'm older and can see through it all
And see where she taught me so I would not fall
She taught me to pray and trust in the Lord

And abide in him and his holy word
I pray to the Lord to bless my Mother
For when she is gone I'll have no other.
So now you see what a Mother can do
She will keep you happy and never blue.
Of course my father was just as good
But it took Mother to raise me to manhood.
I'm thankful, dear Mother, for you today.
That your sorrow and trouble will pass away.
Then when your worldly life is through
God will take you home above the bright blue
There you will reign with him on high
To live in peace and never die.

Your Son, Merle

The above poem was written on Mother's Day May 10, 1942 at Langley Field, Virginia by Private Merle E. Romesberg of the U.S. Army

Chapter 10

A Blind Man and a Black Man and Ghosts

Blackfield Union Church *was founded and called the Blackfield Union Sunday School. It was founded by the following three women: Mrs. Charles W. Fisher, Mrs. Matthew Romesberg, and Mrs. Samuel D. Romesberg. The Constitution is dated May 2, 1920. Presently the church is only used for special services. Photograph by M.E. Romesberg II.*

Across the road from our woods, about one-half mile from our house, was our little church. Every summer Revival Services were held there. For these times, we had guest preachers. At the time of the services in 1933, my two youngest sisters, Betty and Elaine, wondered if the preacher could tell us about the ghosts that traveled back and forth from the church to the grave of Henry Pritts. We were sure that they were ghosts, because our family and friends told us about them. Also, it was rumored that ghosts lived in a deserted house owned by Frank Romesberg. This ghost supposedly dragged a log chain up and down the steps when there was a full moon.

Henry's family had buried him on the farm where he had lived. However, this was not the end of the story. We kept seeing his "ghost"! A light would come out of his grave, cross the field and disappear into the church. This seemed to happen right after dark. Sometimes "it" would come out of the church and disappear at Henry's grave. Each time we saw this, it scared us more than the preceding time.

On Thursday evening, during one of the Revival Services, Elaine and I went to church a little before dark. When we got there, we realized we had forgotten Betty. I remembered seeing her, before it was time to leave, sleeping on the couch. The rest of the family had gone ahead of us. I offered to go back after her, as I knew she would be afraid to come alone, so I headed back home. It was nearly dark when I awakened her and we started back to church. We hurried past Uncle Milt's house, then his barn. So far, so good! There were no lights in the field. We started to run. Then, as we approached the church, a light appeared in front of us. Betty screamed! She turned and started running back toward home. Then another light appeared in front of us! Two lights—two ghosts! We felt trapped—too scared to run! Then one of the "ghosts" spoke: "What are you kids doing out here in the dark? You should be in church!" It was Oscar Poorbaugh. The second "ghost" was one of his boys. They always carried lanterns and as they followed the path from the church into the field along the woods; it seemed to disappear seemingly at Henry's gravesite.

After the mystery of Henry's ghost was solved, we were no longer afraid to go to our little church at night. However, folks kept telling ghost stories, especially on long winter evenings. Finally my brother Merle and Woodrow (Luella's husband) decided to check out the log-chain "ghost" that haunted Frank Romesberg's old house. One night when the moon was full, they hid behind some bushes near the house. The wind started to blow. Then they heard a faint noise. After a while the noise seemed to get louder. Woodrow said, "What are we going to do?" Merle whispered, "Follow me!" and they took off running and didn't stop until they were home! No one ever figured out what made the eerie noise. Later, Frank remodeled the house. No one ever heard the noise again. Maybe it was just the wind blowing against loose boards!

When I was a small boy, the Revival Services were quite boring to me. Pop knew this, so one evening he asked me to recite a poem at the service. "What would that be?" I asked. "Humpty Dumpty," he replied. As little as I was, I knew that would not be appropriate for church, but I didn't say so. Pop stood up and said, "Tonight we have something special. Floyd will recite a poem." I walked up to the podium and stood there looking at all the people—about twenty. I totally forgot what I was supposed to say. So I said nothing and ran back to my seat beside my brother, Paul.

I can remember our Revival Meeting when I was eight years old. Usually I would get sleepy after the worship part of the service and would lay my head on Mom's lap and go to sleep during the sermon. On this Thursday evening, just as I was almost asleep, Paul reached over and slapped me on my leg. I was startled and felt angry toward Paul, when he pointed up to the podium. For the first time in my life I saw a black man. I was wide awake for the rest of the service!

I looked up at this bald-headed man and saw that he was sweating profusely. The sweat was running down his bald head and running off his nose! It had been a very hot day and that evening it must have been 90-95 degrees F in the church. We had made hay that day and I knew how he must have felt. He paused a moment, then pulled out a big white handkerchief from his pocket. After wiping off his face and head, he looked out at the congregation and said, "Friends, I'm so glad I'm a preacher!" He paused briefly and then started preaching. Not for long, however! Again he reached for his handkerchief, mopped his head and face once again and said, "Friends, I'm so glad I'm a preacher!" I don't know how many times he went through this same routine, but I do know that before the evening was over, I had his sermon memorized! "Friends, I'm so glad I'm a preacher!"

The next day was just as hot. We made hay again and my shirt was soon soaked with perspiration. I rubbed the wetness from my face and stood up on the hay in the wagon and looked around me. I looked at the two horses pulling the wagon—at my brothers, Merle, Wilbur, and Paul, forking the hay up into the wagon—the cattle grazing silently in front of a cluster of red oak trees. I looked down the road and saw our weathered barn and the old white house and felt very thankful. I thought to myself, "Hard work and sweat bring about good things."

After we came in with the load of hay at about 4:00 in the afternoon, I went into the house for a dry shirt and a drink of cold water. I heard voices in the kitchen. Aunt Lil (Woodrow's mother) and Mom were talking. Mom said, "Floyd didn't fall asleep at church last night." Aunt Lil replied, "I know and his eyes were as big as saucers.".

The following year, the guest speaker stayed at our home to sleep, eat and visit with the family. I noticed that he had a very big Bible. As he read, he moved his fingers over the pages. Every night at church he sang the same song as a solo.

> Above the bright blue,
> The beautiful blue,
> Jesus is waiting for me and for you.
> Heaven is there not far from our side
> In that beautiful City of Love.

I can remember him saying, "I can see the Lord" frequently as he preached.

This preacher was blind.

Chapter 11

A Message from the Sky

Sketch by Carol Ogle

The "Long Farm" was a great place to go, even though we usually went there to work. There was a long field with woodland on both sides. The field extended from the road, coming from Route 653 to the road that went from our house to Blackfield. Some years we had hay in one part of the field and corn in the other part. Several times a week, after hay making was finished, we let the cows out to graze in the grass. Then we had to watch the cows to keep them out of the corn.

Near the road there had been a house where the Long family had once lived. Their spring was still running and it was a great place to get a drink of cold water, especially on hot summer days.

When the field was ploughed, I often followed behind the harrow or cultivator hunting for Indian Head pennies.

Each Spring we picked horseradish from the field for Mom. One day a man by the name of Hillegas (an itinerant worker) stopped at our farm to work for a few days. He especially enjoyed Mom's good home cooking. At suppertime one day Mom asked him to try some

61

good fresh-ground horseradish. She dipped a big spoonful of horse-radish out of the dish and put it into his mouth. Expecting a real treat, Mr. Hillegas immediately jumped up from his chair and ran outside, gagging and coughing all the way. Everyone laughed but Mr. Hille-gas! Later on he was able to laugh too. Mom pulled pranks like this once in a while! No wonder all of her children liked pulling harmless pranks on people.

I had a path that was a shortcut through the woods from our house to the Long Farm. It was a much more interesting way to go to get there—much more fun than walking along the road. Sometimes I would run and other times I would walk and often stopped to enjoy something new along the way. To take the shortcut, I would start by our red machine shed, where I climbed over the gate and through the orchard and the field behind the orchard, past Henry's grave, through some woods, then through Lesky's field and into our own woods to another section of the field. One day as I ran alongside a fence in the woods, I noticed something white lying on the ground under a green briar bush. I stopped and picked it up. It was a piece of paper about 8x11 inches and as white as snow. It had printing on one side. At the top of the printed side were these words, "The Periodic Table." I was twelve years old at the time and it meant nothing to me. At first I thought someone had dropped it from an airplane. Then I felt that it was something that was probably important. I turned around and ran home to show it to Mom. She didn't say anything, then or later.

At first I thought that this had been a real experience. Days went by and gradually I realized it must have been a dream, but I still felt that it was something important. I felt that some day I would see it again and often wondered what it meant. It would be several years before I would learn the answer.

Chapter 12

a) Housework

Sketch by Emily Morton

Whenever I think of Mom, I think of all the work she did and the old adage:

> Man's work is from sun to sun.
> Woman's work is never done.

With the help of my sisters, Mom cooked three meals a day, 365 days a year. In addition she:

Washed and ironed.

She scrubbed the floors.

She tended the garden.

She cleaned and cooked chickens and wild game.

She sewed, patched and darned the socks.

She doctored the sick.

She baked the bread, many pies, cookies, cakes, and other desserts.

She helped make hay and milked the cows.

As we children grew older, we learned to do many of these chores to help Mom.

All of the cooking centered around Mom's old, black cooking stove. The boys usually did most of the work at keeping coal and wood available. Wood chips and kindling were important to start the fire and sometimes to make the fire hotter.

Mom made the best vegetable soup. She liked to use fresh vegetables from the garden. She cooked what was in season. Her most important vegetable was vine-ripened tomatoes during the summer. She also canned many tomatoes and I think they tasted just as good as those that were vine-ripened when used for cooking. She liked to make a big pot of soup and let it simmer on the back of the stove. This was especially important when our work schedules were such that we often couldn't all come to the table at the same time.

One day E. Jay and I came in to eat at the same time. After eating one bowl of soup, I said, "E. Jay, this is the best soup I've ever eaten." Later, we would ask Mom to make some of that "stuff" we really liked.

While Mom was terrific at cooking three meals a day, she really excelled when cooking for special dinners, including Thanksgiving, Christmas, New Years', Sundays, Threshing Bees, and Butchering Bees.

Fried chicken, mashed potatoes and gravy were her specialty for a home-cooked meal.

Potatoes – Hard firm potatoes were selected. Each was peeled and dropped into a pan of cold water to keep them firm and of good color. They were cooked until soft and then smashed with a hand potato masher. Salt and pepper were added, along with heated milk before mashing was completed. The mashed potatoes were then put into dishes with a big hunk of pure butter and some chopped parsley. Each dish was covered and placed on the back of the heated stove until ready to serve.

Chicken – Fried chicken will be no better than the quality of the chicken used. Our chickens were range-fed, so they could eat as many bugs and grasshoppers as possible. They were also fed oats, wheat, and cracked corn, always with plenty of fresh water and crushed oyster shells. For special events we would use young roosters. On more regular occasions, we used old hens that were left out to forage for several weeks at least before butchering them.

The birds were beheaded and dipped in scalding water, the feathers removed and then cut into "frying size" pieces. Young birds were ready for frying. Older birds were cooked in water before frying. The pieces were browned (with or without batter) in hot lard until done at medium heat. After frying was complete the chicken was placed in covered dishes in a warm oven or on back of the stove.

Gravy – Gravy was made from chicken broth and the drippings from the frying pan with flour and milk being added.

b) Chickens

Sketch by Carol Ogle

We raised chickens for their eggs and for chicken dinners. We traded eggs at Uncle Milts' store for groceries.

We raised White Leghorns mostly because they were good layers. Each hen has seventy-two eggs inside when she starts to lay eggs. Calcium compounds are deposited around each egg. The hen will lay one egg a day in her nest. Then she comes out of her nest and cackles, which is a signal to the rooster to fertilize the next egg before the shell is formed. Laying an egg a day requires a good diet. A high protein meal is used as a supplement with grain. Oyster shells serve two purposes. Chickens have craws in their throats, as do all birds, where the grain is ground to a mash which passes into the stomach. The oyster shells help this process, while also providing calcium carbonate which is necessary to form the egg shells. When you see birds picking along the roadside, they are searching for grains of sand to help digest the hard seeds. Our chickens were allowed to forage where they were able to eat many insects.

Each year in the spring we raised chickens from our eggs. We collected chosen chicken eggs. We placed a board over the nest the hen has chosen, to keep other hens from using it while she gets some exercise, food and water. Some straw is added to the nest to hold the eggs close together.

After the chicks hatched, I put them into a two compartment box which I had built. The box had a flat roof on hinges which was opened to feed and water the hen and the chicks. Fresh clean straw was kept on the floor. The box could be moved to clean ground as needed. With this design (two compartments) the chicks can go through an opening to be close to the hen for warmth. When the chicks were about a week old, they could go through a little door to the outside. The fresh grass was a good healthy place for the chicks. The little door was closed at night to keep out intruders.

I took over caring for the chickens when I was about ten and continued for quite a few years. One year I raised 65 chickens. There were 31 hens which I transferred into a small chicken house for the winter. During that winter, I got up at 5:30 in the mornings. The first thing I had to do was hang a kerosene lantern inside the chicken house. After I cleaned the floor and added fresh straw, I fed and watered the chickens. After I was finished with my chores, I went back to the house and studied. When I was in the eighth grade, I got 31 eggs a day for about two months from 31 chickens.

We usually took eggs to Uncle Milt's store every day and received groceries in return. Often people bought eggs as soon as we got to the store. For their breakfast the next day they would eat one to two eggs! Fresh eggs indeed!

One day after I delivered the eggs, I traded for some oranges which Mom had asked me to get. I left the store with twelve oranges as well as some other groceries she needed. I had a heavy load, so I decided to eat an orange. We had been eager for the orange season as we all liked them. My orange tasted so good, I decided to have another one. When I had nine oranges left I wondered what I'd tell Mom! The more I ate, the better they tasted. Finally I only had six oranges left! I went into the house and put the groceries on the table.

Mom looked in the sack and said, "You only got six oranges?" I made no comment. (Shame on me.)

Chick Brooder Box Floor Plan

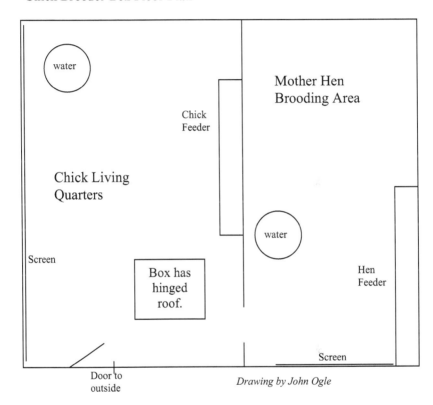

Drawing by John Ogle

c) <u>Cattle</u>

Sketch by Carol Ogle

Many of the daily chores were carried out in the barn. The cows stayed inside during the winter months. They were taken to the outside watering trough twice a day. This job was made easier after we trained our German shepherd, Rex, to herd the cattle to the trough, wait until they were finished drinking, and then bring them back into the barn. If only one person was working, he could be free to clean and bed the stalls while Rex looked after the watering job.

The cows were fed twice a day. In feeding the hay, we had to climb up into the hay mow and pitch a pile of hay onto the threshing floor. Usually we pitched down enough hay to last a week. From the pile on the floor, enough hay for one feeding was dropped through the hay drop (a hole in the floor). Then an amount for each cow was placed in the hay rack in front of the cows' stall.

Each cow was given a helping of "chop" twice a day. Chop was ground corn, oats and wheat, with a soybean supplement added. Another item fed to the cattle, essentially to aid digestion, was chopped pumpkins. A small amount of this was given to each animal. Pumpkins improved milk production during winter months. By the time I was fifteen, corn silage had replaced pumpkins.

A piece of rock salt was placed in the manger for each cow. The cows were milked twice a day. Mom often milked if there was a shortage of help around.

Young cattle, when one year old, were left out to pasture during the spring, summer, and fall. They were also given salt and chopped grain. The milk cows were left out to pasture and brought in twice a day to be milked. We had a bell on one of the cows, so we could locate them. Sometimes they would be in the wooded areas a good distance from the gate. This is where Rex, our German shepherd, made it easy. We would call him to a gate and give him the command, "Go get 'em." Rex always brought all the cows to the right gate!

There were two jobs in the cow barn that I didn't like. Each cow was trained to find her own stall. I hated it when we had to reach around the cow's neck to hook her chain. The job I really disliked was taking our bull to the water trough to drink. We had to lead him with a tethering rod with a short chain on one end and a snap-hook on the other end of the chain. I would open the gate just a little and snap the end of the rod onto the ring in the bull's nose. Then I would have to force him back some into the pen to open the gate. I walked the bull to the trough and waited until he was finished drinking, then guided him back into his pen. We also had a bucket in the pen so we could carry water to him.

Betty, Mom, Wilbur and I did the milking and processing of the milk. For milk customers, we poured the milk into one or more containers and placed them in the water trough in the springhouse. From there it was either picked up by our customers or we delivered it to other customers.

We used a hand-operated skimmer which separated the cream from the milk. We sold cream to the Walker Ice Cream Plant in Somerset. The cream was placed in five or ten gallon metal cans and placed in the water trough in the springhouse. They picked up the cream twice a week.

Mom made butter in the butter churn from cream, which was a tedious job. We turned the handle around and around, a job I didn't like much. Mom saw to it that each of us took our turn.

The water trough where we watered our cows and horses was a place for me to remember. It was intriguing just to see the water pour out of the pipe. I would cup my hands at the end of the pipe and take a drink often, especially if it was a hot day. The water was piped from the Springhouse.

Later Merle rebuilt the Springhouse out of stone. He made a sign to put up over the top of the door reading "Stony Acre Farm."

Taking care of a dairy herd was a never-ending, time consuming job. After Merle came home from the Army in December of 1945, he started to improve the herd and the barn and silo at the former Bittner Farm. He milked cows for twenty-nine years. During this time he missed only two milkings. One time he went to the Penn State School of Agriculture to attend a meeting and another time he went to see the doctor for a cut finger.

We usually butchered two young steers every fall. This would mean two less steers to feed over the winter, as well as providing the family with beef over the winter. Most of it was canned as we had no way to keep it fresh for more than two weeks. Mom liked to make Swiss steak, browned then simmered on the back of the stove for four hours. This made a good meal with little last-minute effort.

Cattle Barn

| Calves | Bull | Cows | Straw Above |

Horse Stalls

Hay Drop

Hay Drop

Hay Drop

Hay

Hay

Threshing Floor

Hay

Drawing by John Ogle

d) <u>Hogs</u>

Sketch by Carol Ogle

We sold some hogs to the hog market; however our main interest in hogs was the smoked meats.

We kept the hogs in a small hog house about 20 ft. x 20 ft. A door opened into a hall which ran about 15 ft. inside the pen. A barrel sat at the far end. In the barrel was hog feed consisting of ground corn, oats, and wheat with soybean meal added.

There were three rooms in the building. Each room had an outside area for exercise and watering. Two of the rooms had an elevated sleeping room.

Each room had a feeding trough. Their main diet was mash and water along with kitchen scraps such as potato peelings, carrot ends, apple cores and left over table food.

One year I experimented with ten young shoats weighing fifty pounds each at the start. I fenced in a three-acre clover field with an electric fence. Each hog got two ears of corn each day. The meat from these hogs, at about 150 pounds each, was very lean. Bacon was especially good.

We butchered three of them, then fed the other seven and sold them at 200 pounds.

The total feed cost was low because they foraged in addition to their regular food.

In other cases we left young hogs to forage in the woods for acorns, worms and grubs.

The birth of little piglets was an exciting experience. One year we had a very large sow. When she was ready to birth, Wilbur stayed up all night to help take care of the little ones. He came to the house at breakfast time and announced that the mother hog had seventeen piglets, all healthy!

Butchering hogs was a big event. After moving to Wilsoncreek, Uncle Alex always came back on butchering day. He liked the job of turning the sausage grinder and the sausage stuffer. John and Luella always drove up from Boswell. John liked to trim the fat out of the sausage meat, while Luella helped in the house.

There were many jobs involved. Early in the morning, we filled two butchering kettles with water and built a hot fire to bring the water to boiling point. Next, we shot and bled a hog. Boiling water was dipped out of the kettles and dumped into the scalding barrel. The hog was lifted onto the table and two men slid one end of the hog into the hot water. They pulled in and out, rotating it at the same time.

After about six dunkings, the hog was turned end-for-end and the process was repeated. The scalded hog was then pulled out and onto the table. Two to four men scraped off the hair and dirt. While the scraping was being done, someone refilled the kettles with water and built a hot fire for the hot water to scald the next hog. In the next step, the hog was gutted, hung on a three-legged scaffold, then split into two pieces.

Next, the head, heart, liver and the two front feet were removed and held in a cool place to be cooked later for Mom to make "ponhaus," a delicious breakfast food. The small intestines were taken to Mom. She turned them inside out, scraped and washed them to be used for casing for sausage.

In the next step, the two sides were cut and trimmed for smoking. The following cuts were made from each hog:

> Two hams
> Two shoulders
> Bacon
> Backbone for roasting and pork chops
> Fat was trimmed off for lard
> Lean meat for sausage

After the first hog was cut up, several people started to trim fat for lard and meat for sausage. The sausage meat was ground. Everyone enjoyed the next step. After some ground sausage was put in a big pan, Pop added salt and pepper to sample. He fried some fresh sausage which everyone enjoyed.

Links of sausage were made with the aid of the "sausage stuffer." The stuffer consisted of a heavy metal container, about two gallons in size, with an opening at the bottom and a lid attached to a screw, which was driven by a crank. The crank forced the sausage meat out the bottom hole and through a short pipe. The casing was fitted over the outside of the pipe. As the sausage was forced out of the pipe, it pulled the casing off the pipe, forming it into links.

At the end of the day, the fat was rendered and the meat for Ponhaus was cooked. All the fat was placed in a big kettle (or two if needed). A gentle fire was built for this step. Someone needed to stay close at this point. If froth started to rise in the kettle, we cooled the fire and, if necessary, a spoonful of butter was added to break the froth.

After the fat cooked for a designated time, the fire was reduced and the hot lard was poured through a sausage stuffer, using a plate with holes in the bottom with a filter cloth to hold the cake of rendered fat. This was called "cracklings" which are quite tasty, especially while still hot. The lard was collected in gallon crocks.

To make Ponhaus, Mom separated the bones, skin, fat and gristle from the broth and meat. Then the meat was ground. The following recipe was used to make Ponhaus. This recipe was tracked down for me by Violet, my brother Paul's wife.

1 lb. chopped cooked meat

2½ tsp. salt

¼ tsp. pepper

2 cups corn meal (or a mixture of wheat and buckwheat flour)

2 med. chopped onions, browned slowly in lard

2 ½ quarts broth

The meat, seasonings and onions were mixed and added to the cornmeal and simmered for one hour in the broth. Then the mixture was poured into pans and cooled. It was then ready to slice and fry, and serve with maple syrup.

We liked to cook a big piece of backbone with sauerkraut, to be served with mashed potatoes and fresh baked bread slathered with butter and topped with apple butter.

Pop laid out the hams, shoulders and bacon on a big table in the Summer House. He then rubbed a thick layer of salt and pepper on each piece. They were left on the table for two days before being taken to the smokehouse. A small hole was cut into each piece of meat to accommodate a cotton cord. A loop was tied at the end of the cord in order to hang the meat on spikes pounded into the ceiling or shelves. The sausage links were wound around slats of wood, suspended from the ceiling.

I thought it was fun taking care of the smokehouse. The first step was to gather an ample supply of green maple and hickory. I usually cut small trees and small limbs from bigger trees. Pieces were cut in lengths to go into the smokehouse stove. I liked to build a hot fire out of dry wood. After the fire burned down to hot coals, I'd fill the stove with green wood and reduce the draft. The fire would then smolder for about four hours. As the fire burned, I'd add more green wood.

After about three weeks, the smoked meat was ready to go into storage. As long as we had cold weather, we stored some of the meat in the attic of our house. The best place to store smoked meat was to bury it in the oat bins in the granary. Mice couldn't get to the buried meat without suffocating. We kept this meat buried from October until June.

Everyone liked Pop's smoked hams. At our home we had ham dinners, ham and eggs and ham sandwiches. One day a big shiny new car drove up and a well-dressed man stepped out of the car. He was from West Virginia. He said to Pop, "I hear you smoke a good ham. Would you sell me one?" Pop never said a word, but headed toward the granary. The man was surprised to see Pop come out of the granary with a big ham in one hand.

One day Pop asked me to go with him to check on a forest fire which had occurred the day before. Pop was the Fire Warden for the area. I carried a big lunch pail and two fabric buckets for two miles along the dinky track back to the river hill. Then Pop said, "Bad news! We're going to have to back-burn." A section of the front was still smoldering. This meant hard work—raking leaves and carrying water for quite some time. We finally finished about 2:30 P.M. and were ready for a rest and something to eat. We opened the lunch bucket and found that Mom had packed two big slices of smoked ham along with shanty potatoes—a reward worth all the work!

e) Horses

Sketch by Carol Ogle

Our horses on the farm were big draft horses. They were our only source of power for pulling farm equipment, as we didn't own a tractor until 1940. Up until then, all of the farm equipment was pulled by our two horses.

The wagon was used all year for different jobs. During the haymaking season, it was fitted with a hay rack. The remainder of the year it was fitted with a box. Some of the items hauled with the box attached included mining posts, coal, firewood, brush and logs. During the harvest season it was used to haul potatoes, corn and grains.

Winter was a good time to drag logs to the sawmill. Several times after a deep snow, I can remember Pop getting up early and dragging a log down the road to school. It was then a nice, easy walk for us, compared to walking in the deep snow.

When the snow was deep, the bigger boys always went to school first, so that when the younger ones went, it was easier walking in their path.

Plowing was probably one of the most important and demanding jobs for a team of horses. Two horses and one man with a single-bottom plow could plow about one acre a day. We had 65 tillable acres and plowed about 40 acres. The remaining acreage was used for hay-making. This would limit the size of a farm with only one team to about 80 acres.

After plowing the fields, the next step was harrowing. Heavy clay soil can be difficult to harrow because it can be too dry and hard lumps will form. If the soil is too wet, clay balls form. The goal is for the soil to have a finely divided texture. Sometimes we had to harrow three to four times, until the soil would be suitable for planting. Seeds germinate more quickly and the plants grow faster in fine soil.

A "corn planter" is used to plant corn and a "drill" is used to sow oats, soybeans and wheat. We sowed grass seed with a hand-cranked spreader for planting a hay field in the early years.

Our team of horses was used to pull the binder to cut oats and wheat when ready for harvest. We also used a corn binder to cut corn for silage and also to cut the ripe corn to be shucked and later the corn removed from the cob for feeding hogs and cattle.

I had two run-away rides on horses as a child. I was about eight when Wilbur was plowing the Long Farm. He told me to ride the lead horse, Maude, and lead Frank on my right with a rein. We were going home for lunch. We had to go on a road in the woods to get to the main road. There were many limbs reaching over the road, and I was having a time dodging the limbs. The horses were moving at a fast pace. Likely, they were anxious to get home for a good watering. Suddenly, Maude bolted! I was afraid the limbs would drag me off of her back, so I let go of both reins, grabbed the shoulder pad and laid forward. I felt myself bouncing on her back. We approached the gate to the main road. I thought she was going to run right through the gate. I kept yelling "Whoa Maude! Whoa!" She banged against the top rail and stopped. Wilbur was soon there and grabbed the reins. I hit the ground; both feet first and said to Wilbur, "You take them now!" Wilbur said, "I think a bee must have stung her!"

My second episode with a run-away horse occurred when I was riding Frank to unload hay at the Bittner Farm. Paul was up in the haymow. Merle was handling the fork on the wagon and Wilbur was watching me while attaching and unhooking the rope from Frank as I drove back and forth for each load of hay. Just as Wilbur attached the rope and Merle hooked the fork in the hay, Frank bolted! He ran until the load on the fork hit the barn roof. At that instant, Frank went down head first and I was ejected into space! I think I finally landed on all "fours." Frank had broken his harness and kept running, past the house, through a barbed wire fence into the field. He broke four strands of barbed wire fencing! Halfway through the field, he stopped and ran right back to me. At the time, I thought he had come back to see if I had been injured.

Daily care of our horses was an important chore. We fed them twice daily and walked them to the watering trough twice a day. We gave them a thorough combing and brushing twice a week.

We were very affectionate toward our horses. One time a horse named "Babe" broke a leg. Pop made all the kids leave while he took care of the horse. Pop cried when he shot him.

Another time we had a big red horse named "Red" that became very ill and was close to death. The family was about to go to the Romesberg reunion. I asked Mom if I could stay home. She said "yes." After the family left, I hurried down to the barnyard to be with "Red." He had been down all night. I went to him and stroked his head and neck. I could see where he had scraped the ground with his legs and feet, trying to get up. Now his legs were still. I stroked his head one last time. Then I climbed up on the gate and watched him. Finally I could no longer hear him breathing. I sat there for a long time. After a while I headed for the woods and a long walk.

f) <u>Oats and Wheat</u>

Sketch by Carol Ogle

There were many early American scenes that touched the hearts of travelers as they traveled by the farms. Fields of oats and wheat provided many such scenes, especially the wheat.

As the grass-like plants grew, they cast a deep green glow of unequaled beauty over the rolling hills and valleys. As the stalks grow taller, they gradually turn from green to golden. The head of the stalk develops into a cluster of seeds. Each seed is encased in a shell and each shell extends into a needle-like spike.

The field seems to turn into a huge carpet. As gentle breezes blow, stalks bend in one direction, nearly touching the ground. Then the stalks straighten and bend in the other direction. Each stalk swings back and forth. This motion spreads across the field and across hill and valley like gentle waves on a great ocean.

Travelers drive by slowly to enjoy the view from their car windows. Some might stop and take a deep breath, in awe of nature's beauty. This site is witnessed across the U.S., especially in the midwestern states. In a short two to three weeks this "moving" scene changes to a more "stationary" one, encompassing the many shocks of grain scattered across the fields.

Up close, these shocks look like little "huts." Farther away, they look like little dots between the field and the deep blue sky. Today, we no longer see shocks of wheat because of modern farming methods, specifically the use of combines. Today, we can still see the golden fields of ripe wheat and oats. In fact the fields are usually larger. Before the invention of the combine, farmers used binders to harvest grain. Bundles of wheat, oats and corn were stacked outside to allow the grain to dry in the sun. Today the farmers use large tractors to plant larger fields. The grain is allowed to ripen while standing, as much as possible. Then the grain is cut and threshed in one step by the combine. The grain is then hauled to a central location where it is dried and stored.

Horses, binders and threshing machines are a thing of the past. The need for shocking the grain in bundles has been eliminated.

Threshing in Black Township[12]
The old steam engine and thresher went from farm to farm. It was owned and operated by Sam Kretchman, and shown on his farm, which was later owned by George Pritts.

Binding Grain on Hauger Road[12]
This is on the William H. Weimer farm in the early 1920s.

g) <u>Corn</u>

Photo by John Ogle

Corn is one of the most important crops grown in the United States. It was also an important crop grown by Native Americans. We first learned about corn from the Indians during the early colonial days. Usually, they had small fields which they hoed and smoothed before planting. They also put a small fish in the ground where each seed or group of three seeds was to be planted. The planting area was often edged with stones, which the Indians had thrown out of the field before planting to make it easier to work the soil.

In the early years of farming, small plots of sweet corn were managed with manual labor. The soil was turned over with a spade, and then worked with a garden rake. Furrows were made with a hoe and seeds were dropped in the furrows by hand. The hoe was then used to cover the seed with about an inch of soil.

During the growing season, a hoe was used to cut out weeds and grass and to "hill" ground around the plants. For larger fields, horses were used to plow, harrow, plant and cultivate. Sweet corn was harvested while the kernels were still creamy inside.

Pop used to tell us that corn should be planted when the white oak tree leaves were the size of a squirrel's ear. Mom always said corn should be knee high by the Fourth of July.

In the fall, generally after the first frost, the corn stalks were cut by hand with a sickle. The stalks were placed in an upright position against a three-legged corn buck. When enough stalks to make a shock were standing against the buck, the shock was tied with a piece of binder twine and the buck was removed.

We often planted pumpkins in some of the cornfields. We would remember the old saying, "It's time to husk the corn when the frost is on the pumpkins and the fodder is on the shock." Actually, corn husking began after the sun and wind had sufficiently dried the kernels of corn.

On Saturdays when I was able to help with husking the corn, we would usually have five people working. We would pull the shocks in a circle of five workers. Each person kneeled on the outside of his shock. A hand tool (a knife-like tool fitted on a glove) was used to peel the husk from the ear. The ear was then broken from the stalk and thrown on a pile in the center of the circle.

Later the corn was hauled to the corn crib and unloaded with a shovel. The crib was narrow at the bottom, about four feet, and wider at the top. The construction included a roof with the sides made of three inch-wide slats, separated about one inch. This allowed air to circulate and kept the corn from molding.

I took agriculture classes my first two years in high school. Both years, our school held corn husking contests. The first year I competed, I figured Charlie Kaufman, a senior who was not only older but also stronger than me, might win. I wanted to win, so I stationed myself so I could see how fast he threw the ears onto his pile. That way I could tell how well I was doing compared to Charlie and I set out to beat him. I noticed that he threw the ears as fast as I did, but he wasn't steady. By the end of the allotted time, I had beaten him by one-half bushel! I won the contest again the following year.

Sweet corn was one of our favorite vegetables. Small plots of sweet corn required frequent hoeing to keep out the weeds and ensure a good crop. We did not enjoy hoeing corn! Sometimes we took advantage of fires from burning brush during sweet corn season. We roasted corn with husks on in the hot coals.

As an adult, my family attended a church retreat at Green Lake, Wisconsin one year. We had a corn roast for a group of about seventy people. There was a pile of oak logs available. I started a fire soon after daybreak. By the end of the day there was a pile of hot coals and ashes about ten inches deep by five feet in diameter. The corn was picked in the late morning. The husks were pulled back and each ear was covered with butter and salt. Then the ears were dipped in cold water. Then each ear was wrapped in aluminum foil and placed on about two inches of hot charcoal and covered by about four inches of burning embers. After about thirty minutes we started the feast. All the ears were removed from the coals in less than forty-five minutes. They were cooked "just right" and the corn roast was a great success!

h) Hay

John Ogle

Hay is essential for horses and cows. We planted a mixture of red clover and timothy. Each year there seemed to be less red clover, so our hay became mostly timothy. The red clover-timothy seed mixture was planted with the oats using a shoulder seeder.

We cut hay when dry weather was forecast. The hay was cut and left to dry for about forty-eight hours, hoping it wouldn't rain until the hay would be safely in the barn! Then it was raked into rows, gathered with a fork and loaded into the hay wagon. After hauling it to the barn, we unloaded it by hand into the haymows.

The biggest advancement in hay making equipment was the invention of the baler. The baler picked up the hay in the field and compacted it into a rectangular shape that weighed about sixty pounds. These bales were ejected out of the bailer onto the ground, or directly into the hay wagon, ready to be hauled to the barn. There, the bales were loaded onto a conveyor which carried them up to the haymow. It was easier to feed the cattle with baled hay rather than loose hay. It is very important to be sure the hay is dry, before storing in large quantities. Wet hay can cause mold as well as undergo spontaneous combustion and cause a barn fire.

i) Potatoes

Photo by John Ogle

Picking potatoes in the fall was a lot of fun, especially when there were really big ones. The potato season started early in the spring. The first step was the potato "Cutting Bee." We used leftover potatoes from the winter. Small potatoes were planted "as is." Larger ones were cut into pieces, with each piece having at least one eye, preferably two.

The ground was plowed as soon as the field was dry enough. It was harrowed several times, to kill as much grass and weeds as possible. A ditch (or furrow) about ten inches deep was made with a one-horse plow. The potato pieces were then laid by hand, eye up, in the bottom of the furrow.

We liked to cut the potatoes two days before planting. This allowed the new cut to dry, so they would not rot before sprouting and growth began. The planted potatoes were covered with soil by hand, using a hoe.

Fertilizer high in potassium and phosphates was used when we had money to buy it. Any wood ashes available were also used. Potatoes for winter use were dug on dry days to keep from bringing too much soil into the potato cellar. Potatoes for winter were best planted in late spring (around May 15) and were dug in the fall as late as possible.

We dug the potatoes with a furrowing plow, a little wider than the one we used for other crops. During the growing season we hoed out weeds, and "hilled" the soil around each plant. We used a horse-drawn cultivator several times during the summer.

When I was about ten years old, we bought a used "one-row" planter and a "one-row" digger. The digger plowed under the vines and the potatoes. Both were lifted onto a conveyor made of rods. This belt-like chain shook the soil off first and then spread the potatoes in a row on top of the ground. As the potatoes were scattered ahead on the ground, the pickers followed with buckets. We children picked and filled our buckets, and were handed another empty one, while one of our older siblings dumped the filled buckets into the wagon. When the wagon was full, we'd run to the wagon, hoping for a ride on the wagon seat. When we reached the house, the potatoes were carefully carried down to a dark corner in the cool cellar. During the winter days I took my turn answering Mom's call, "Bring up a pan of potatoes for supper."

I can remember one year when Merle was planting potatoes and ran out of fertilizer. He gave both Paul and me each a three-gallon bucket and told us to go to the place where we had burned brush and scrape out as many wood ashes as we could. Merle dumped both buckets of wood ashes into the fertilizer box. We marked the place where we started with the wood ashes with two sticks. Where we used the ashes, the vines were bigger and greener and they lived longer. Most important we had double the yields of bigger and better look-ing potatoes!

Potatoes were a mainstream in our meals. We had them for break-fast, lunch and dinner. Mom cooked fried potatoes, mashed potatoes, "shanty" potatoes, salt-water potatoes, scalloped potatoes, potato salad and creamed potatoes.

For breakfast, Mom rotated different foods such as cold cereal, hot cereal, bacon, eggs and fried potatoes. No matter what the dish, she always peeled the potatoes and put them in a pan of cold water to keep them fresh and of good color until ready for cooking. *"Shanty" potatoes* were sliced before they were cooked in salt water and butter added. *Creamed potatoes* were sliced or cut up in chunks and cooked with milk and cornstarch. *Saltwater potatoes* were cooked whole until soft. They were served and mashed at the table with butter, salt and pepper, or covered with gravy.

When I was about five years old, Merle asked me to go along with him to some neighbors who were needy. It was fun riding on the old wagon seat right behind the horses. There were many children in the family and Pop was often generous with them. Each spring we gave them our "left-over" potatoes. When we had electricity installed, he gave them our carbide light system for their house.

j) Apples and Cherries

Photo by John Ogle

Matt Romesberg filled a fifty-five-gallon barrel with apples for winter use. He kept watch on the apples and removed any that showed signs of rotting. He cut out the bad part and ate the good part. He continued doing this until the barrel was empty. Thus, he succeeded in eating a barrel of rotten apples!

For a number of years after I was ten, I helped pick and store apples in the garden for winter use. First, we spread a layer of straw on the ground, about six feet in diameter. Next we spread a layer of apples. Then we alternated layers into a dome-like structure, until it was nearly four feet high.

At this point, we shoveled a layer of ground over the pile, and then dug a drainage ditch around the pile. We did not use these apples until all the apples stored in the basement had been eaten. Usually the pile was covered by a foot of snow by the time we opened a hole, burrowing through the snow, ground and straw. This was typically about December 1. We would take a bucket of apples from the pile each week. Red Delicious, Winesap and Northern Spy were our favorite apples to bury for winter storage. Our buried apples usually lasted until April. We wouldn't have apples again until July, when our early apples would be ready.

We used a ladder and bucket to pick the apples, being careful not to damage the surface of each apple. I always used cotton gloves to prevent damage to the natural wax layer.

Some years, Pop grafted apple trees. Some of our trees bore poor quality fruit and some of them were volunteers grown from seeds. Pop took scions (living sprouts) from good apple trees and top-grafted them onto a poor-quality tree. However, we did have some trees grown from seed, which grew especially flavorful fruit.

There were some apple trees growing in the fence row along the road to Harrison Pritts' house. One of these trees produced very early apples. It was a greenish apple with a vivid red flush. When they ripened, I would check for any that fell on the ground, before bringing the cows past the tree. They liked these apples as much as I did!

After watching Pop grafting, I started doing some of this myself when I was about ten years old. I liked to pick a favorite apple and graft the scions onto a semi-dwarf root stock.

Many people have flowering crab apple trees. Some root growth grows many "suckers" which need to be cut off. Occasionally, these "suckers" are planted and can produce a large tree with different blossoms, which are less desirable. On the other hand, one can produce a beautiful flowering crab or a semi-dwarf apple tree. These "sucker-trees" can be used as a base for any kind of apple trees, as well as flowering crabs.

In 1953, when living in Michigan, I bought a semi-dwarf, Cortland apple. By 1965, when we moved to a larger home, we were getting two bushels of really good apples. Two years later, I went back to our old house in Michigan and rang the bell. I asked the lady if I could get some of the sucker shoots from her apple tree. She replied, "You must be Floyd. I want to thank you for the beautiful apple tree!" Then she went to the basement and brought up a bag of frozen sliced apples. She said, "Look how white they are." She had used citric acid to keep the peeled and sliced applies from turning brown. I have moved three times since then, repeating the grafting process, and still have an example of the reproduced Cortland apple tree.

Every year back on the farm, we picked bags of apples and hauled them to the apple-butter factory to be processed. Mom canned this apple butter, which we enjoyed all year. We also had apple cider made at the same place.

John and Luella came often on weekends. Mom always baked fresh bread. As soon as they arrived, she served them some bread, so that John could have apple-butter bread and cold milk. This was a great favorite of his!

We had a little fun with a barrel of cider. Betty and I were the culprits. We cut a piece of wheat straw, long enough to reach down to the cider. So we drank cider out of a thirty-three-gallon barrel. We kept the stopper on tight to slow down fermentation, but eventually the cider turned to vinegar. We all enjoyed the cider, but Mom then had vinegar to use all year.

In addition to storing apples, Mom also dried and canned them.

We had another prize fruit tree, besides apples. These trees were in or near fencerows in all directions around the house. In the spring, they were covered with beautiful white blossoms. By July 15th, they were covered with lovely, green leaves, along with fruit. People sought after them. We shared them with our neighbors. They were our favorite fruit. These were our sweet black cherries!

We picked gallons and gallons of these delicious cherries. Mom canned them for winter use. We ate them for dessert, seeding them as we ate. Mom seeded them by hand for making pies.

To pick the cherries, we used gallon buckets, ladder and a "cherry hook." An "S" shaped wire hook was used to hang the bucket on a tree limb. We used a "cherry hook" to pull branches filled with cherries in close so we could reach them.

The day I stayed home from our Romesberg reunion was the day our big red horse was dying. I stayed with him until he died. Then feeling sad, I took a long walk in the woods. I circled through our three wooded areas. On the way home, I saw a grey squirrel building a nest. I sat down and watched. It would bite off a small branch with leaves and tuck it into the nest. After a while, I walked on past Harrison Pritts' house and realized I was hungry. I crossed the field and headed for the cherry trees. I pulled down a low limb and began to grab handfuls of the delights. I gorged on the cherries and went back to the house to wait for my family to come home. Mom wasn't too surprised when I said I wasn't hungry. She could tell from the red stains on my face and hands!

When I was a little older, about thriteen, I noticed a tree growing in a fence row behind our garden. The tree was fairly small and the fruit was golden-yellow and quite large. It was slightly tart, but very sweet. I wondered if it had developed through the cross-pollination of our sweet black cherries and our red pie cherries. The year I left Bucknell (1950) to attend the University of Cincinnati, Jenny and I came home for a few days. I picked about ten gallons of these delicious yellow-golden cherries. Mom canned twenty quarts for us to take to Cincinnati. I finally concluded that this "hybrid" was unique. However, before I could do any further investigating and by the next summer, the tree had been destroyed. I have often regretted that I failed to act in time.

k) Hard Jobs

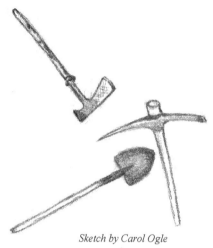

Sketch by Carol Ogle

We continued clearing land until about 1935. This job involved removal of trees and stones. Trees were cut for lumber and mining posts. The most strenuous job was one man cutting down a tree (especially a large tree) with a one-man logging saw. Usually this job was done by two men with a two-man logging saw. These saws had about four teeth per inch. Removing large rocks and stumps was hard work. We used dynamite on the most difficult jobs. It was difficult work to dig ground with pick and shovel and cut roots with an axe. Trees with large tap roots were especially difficult.

Much of our ground was heavy clay. Low areas stayed wet late into spring. We dug ditches to drain these wet areas and allow the ground to dry out for planting. The hardest job was digging the ditch two feet deep. Sometimes the ditch was quite long in order to drain the water into the woods or to a ditch along the road.

We had a lot of hard work removing animal manure from the barn; especially from the pens where the young cattle were kept. Straw and manure were compacted, often two feet deep, and difficult to dig out with a pitch fork.

It was critical to keep axes and saws sharp. Pop sharpened the axes with an old-fashioned grindstone. We kids took turns turning the wheel with a hand crank, pouring water on the wheel to keep it clean.

Pop used the following steps to sharpen saws:

1. File cutting edge so all of the teeth have the same height,
2. Use a saw setting tool to set the girth of each tooth. This step was determined with a cut,
3. Sharpen each tooth.

l) Gardening

Photo by John Ogle

Gardening was another important job on our farm. We had a garden about 100 feet by 100 feet near the house. Mom was the chief gardener when I was little. Wilbur did the hard work for her and when I was about eleven, I helped a lot more. We worked the soil with a spade shovel, garden rake and hoe. We divided the garden area into beds. We used the spade to turn the soil and make a ditch. I shoveled the ground out of the ditch, and then filled the bottom of the ditch with manure. The ground out of the second ditch was thrown on top of the manure. After four rows, the ground was raked with a garden rake, pulling the ground forward to fill the last row of manure.

We planted just about everything; lettuce, radishes, green onions, peas (shell), green beans, tomatoes, cabbage, sweet corn, beets, peppers, carrots, sweet peas, cucumbers and squash. We also planted a large area in one of the fields. We called this our truck patch. We planted a large amount of cabbage for winter storage and for sauerkraut. We planted tomatoes and sweet corn and green beans. Some of the beans and corn were both canned and dried. Mom did the drying in a big flat double pan with water in the bottom and corn or beans in the top.

Beans were planted and allowed to ripen and then shelled! We planted navy, Lima, kidney and horticulture beans.

Cucumbers were grown for pickles. We also planted pumpkins, turnips and rutabaga.

Cabbage, tomato and pepper plants were started indoors, and then transplanted. We watered all plants that were transplanted. We carried the water with buckets and poured it on the plants with a tin can.

The best part of vegetable gardening was meal time in the winter. We always had three or four vegetables at each meal. Mom really knew how to prepare and serve vegetables.

Chapter 13

School Days and WW II

Sketch by Carol Ogle

Maple Syrup

Someone once said, "Life is the sum of all the moments that pass and history is the sum of those moments that are recorded." The Depression lasted from 1929 to about 1940 and the beginning of WW II. I was born in 1927, so I lived through the Depression and the war. My father didn't make much money. He spent almost all of his time farming and working in the woods. He had a job cutting brush under electric power lines from 1936 to 1940 for $.65 an hour. He also was a fire warden and the constable in Black Township for many years. We spent very little money. New clothing was a luxury. Mom patched and mended our clothing, and outgrown clothing was handed down to a younger child.

However, we never went hungry. We grew corn, wheat, oats and hay. We raised cattle, hogs and chickens. We traded eggs for groceries at Uncle Milt's store. We had a big garden and mom and the girls canned many quarts of tomatoes, green beans, beets, pickles, apples, cherries and beef.

We were taught to work when we were very young. As soon as we could carry a small bucket, we helped water the garden plants. As we got a little older, we learned to pick cherries, apples and pears from the fruit trees. We also picked tomatoes and dug up potatoes.

I can remember my first day of school. Betty had told me that if I had to use the bathroom, I should hold up my hand and say, "May I please use the bathroom?" This meant that I would have to walk past the big kids and then go outside and hunt for the boy's outhouse. Before I could get up enough nerve to raise my hand, it was too late! I had wet my pants on the seat and the floor! When the bell rang, I got out of there and ran home in a hurry.

I can remember showing my report card to Mom when I was in the third grade. Mom said, "What is this grade in Spelling? What homework do you have?" I showed her my *fourth* grade spelling book. The next day I marched to school with a note for the teacher in my pocket saying, "Please use the *third* grade spelling book for Floyd," signed Mrs. E. J. Romesberg.

What I remember best about grade school was my friends: Norman Hay, Bill Shaw, Charles (Sonny) Weimer, Lyle Walker, George and Bob Mitchell and Frank Omerzo.

Softball was very important to us from fourth grade through the eighth. A league had formed with four schools participating: Wilsoncreek, Blackfield, Wable and Humbert. When I was in the fourth grade, I became the pitcher. It was our first game. I threw the first pitch without warming up. The ball sailed 25 feet above the batter's head. One of the Snyder twins on the Wable team said, "This is going to be easy!" They gave me three pitches to warm up. Then I promptly struck out three batters in a row! I pitched for five years for Wilsoncreek and we lost only one game.

All the kids in my family attended Wilsoncreek for eight years. We lived about one mile from the school, so it was an easy walk, but not so easy in the winter with snow on the ground.

One of the consequences of WWII was "Rationing." Sugar was scarce, so maple syrup was in demand. One year I had helped Wilbur make maple syrup and learned how. So in 1940 Betty and I decided to make some maple syrup for Mom. As far as I knew, we had only one sugar maple tree. It took forty gallons of sap to make one gallon of syrup. We had northern maples, but the sugar content in the sap was lower, so that it required sixty gallons of sap to make one gallon of syrup.

The first step necessary to make syrup was to make "spiles." Spiles were made from the stalk of an elderberry bush about three-quarters of an inch in diameter. The stalk was cut into eight-inch lengths, then about one-half of the circumference was cut off about one-half up the length. Next a wire was used to push the pith out from the center of the spile. The spile was then tapered, so that it could be tapped into a hole, bored into the tree. The most difficult task at this point was finding enough one or two gallon containers. We used milk buckets, water buckets and pots and pans.

We collected fifty containers and made 160 spiles. Each container was placed at the base of a tree. The spiles were placed about two or three feet off the ground, so that the sap would drop into the buckets.

Tree Diameter	Number of Containers per Tree	Spiles per Tree
1 foot	1	2
2 feet	2	4
3 feet	3	6
4 feet	4	8

We mounted a 30-gallon barrel on a sled pulled by one horse to collect and store the sap. This was taken to the sugar house, where we had two iron kettles which we filled with sap and brought to a boil over a hot wood fire.

We started tapping the trees in late February. Cold nights (below 32 degrees) and warm days made the sap flow and gave good runs. Betty and I collected about 750 gallons of sap that year. We kept carrying wood to feed the fire and keep the sap boiling. We kept the smaller kettle nearly full by dipping the hot sap from the larger kettle. We kept adding cold sap from the barrel to keep the large kettle full.

When we needed to get more sap, we emptied the barrel into a wash tub. We boiled the sap until about two gallons of concentrated sap would give one gallon of syrup. Then we would take it to Mom to finish the boiling process on our wood-burning stove in the kitchen. I usually kept the sugar house fire going until midnight.

The season lasted about five weeks. We made twelve gallons of maple syrup that year. Pancakes with maple syrup were enjoyed by the family for a long time.

When it was time for high school, the girls, Luella, Helen, Betty and Elaine, walked the four and one-half miles to and from Rockwood High School. They all graduated. It was a different story for the boys. Merle went for eight months, then quit to help Pop with the farm work. Wilbur went for one month and quit for two reasons. One was to help with the farm work and the other because he would have had to wear Pop's old band uniform if he went to school. There was no money for school clothes. Paul registered to go to high school. I can remember him proudly telling me that he was going to play Pop's trumpet in the school band. He didn't go even one day. I wished he would have gone. He could have been a star basketball player as he was very tall!

As for me, I didn't do anything about going to high school. Then came Mr. Forbes along with my sister Betty! Mr. Forbes was our new agriculture teacher. That summer of 1941, Mr. Forbes drove his new car to our farm. He was recruiting students for his fall classes. We had just pulled into the barn with a load of hay. He walked up to the wagon and called out, "Are you going to go to high school?"

I answered, "No, I don't think so."

He asked, "What are you going to do?"

I told him I would be staying home to work on the farm. He left, but came back three times and talked to Pop each time. Then I can remember my sister Betty having a somewhat heated discussion with Mom and Pop on the matter. Betty told them there wasn't room on the farm for another brother. She also reminded them that I had good grades while at Wilsoncreek. I studied every morning at 6:00 A.M., after attending to the chickens. The high school wanted me to play softball and basketball. Nothing further seemed to happen.

Then the day before school was to start, Mom said, "Here are two shirts and a pair of pants."

Pop said, "You are to go to high school!"

The next morning I finished my chores early, had breakfast and was ready for my first trip to Rockwood High School by 6:00 A.M., along with my sister Elaine.

As we passed the trees and big rocks along the way, I kept thinking about what the future would bring for me. Then it came to me. Mom and Pop were watching me. They wanted me to decide. Would I stay on the farm or continue this four and one-half mile walk to high school every day and figure out what my future would be from there? I felt a little tingle down my spine. My back was turned to the farm, the trees and big rocks, and my face was turned toward the future. It was my decision! I would go to high school and do my best in my classes, as well as softball and basketball.

We arrived at school early that first day. I went to the principal's office and he gave me a schedule for my classes. The first event I can remember was softball practice. After two practice sessions, the coach told me I would be playing shortstop and be the number two pitcher. We had a game scheduled for the second week. The day of that first game there was an announcement. Softball would be cancelled for the year because of the polio epidemic. Then, because of the war, softball was cancelled permanently.

One week later I touched a basketball for the first time. I traveled with the junior varsity team my freshman year, except for the first game, and by the last game I was on the varsity squad.

The war with Japan broke out just after I started high school. Gradually, most of our men teachers left to serve in the armed forces. During those war years, I can remember the family gathering around the radio to hear the latest news. We also looked forward to Merle's letters from India where he was serving as a supply sergeant.

I continued with my classes and then, in my junior year, I attended my first chemistry class and had a memorable experience. As our teacher handed out the chemistry books, I took mine and placed it on my desk. Then I opened the cover. I froze and stared at the print.

There for the second time I saw the words THE PERIODIC TABLE! The first time was in my dream, when I found a copy of the "Table" under a briar bush along the path to the Long Farm. I was ten years old at the time. I felt that was a message for me. It was two-fold.

1) I should go to college and study chemistry,
2) I should learn as much as possible about the table.

I then went to Mr. Forbes, who had become a mentor for me, to ask if it was possible for a person like me with no money to go to a college like Penn State. He assured me that it was possible. His advice was to work and save the money, both during the summers and while in college. At that point, I made a firm decision. I never looked back or had any doubts about this decision.

High school events became somewhat routine:

1) Walking four and one-half miles to and from school, rain, shine or snow,
2) Studying,
3) Doing my share of the farm chores,
4) Basketball.

When we had basketball games at home, I walked home after school, did my chores, had dinner and walked back to school for the games. Walking home after the games was challenging, especially after an "away" game. Then I wouldn't get home until between eleven and twelve P.M.. Mom did not go to bed until I came in the door.

One morning it was very cold. Elaine and I had walked about one and one-half miles, when Elmer Pritts picked us up in his dump truck. That day the principal called an assembly for all the students. He said that he would like to applaud some students for their efforts to make it to school during bad weather. He asked six students to stand up. Elaine and I were two of the six. It was -26 F that morning!

I didn't have to walk to school all the time, however. My brother Wilbur bought a bike for me, which I rode in good weather. I had one very scary incident with my bike. Elaine usually left for school before me, and if she was late, I would pick her up on my bike. There was a very steep hill with two right angle curves down across the wooden bridge which crossed Coxes Creek and the railroad. One morning the brakes on my bike grabbed and gave out. We were just starting down the first hill. I told Elaine, "The brakes are gone. When I say lean, lean!"

We banged onto a loose plank as we reached the first bridge and came to a 90 degree curve just past the bridge. No car came toward us! We cut clear to the inside edge of the road and stayed in the oncoming lane. Then we crossed back to our rightful lane, cutting the "corner" on the second curve. We ran through the stop sign and coasted down East Main Street. I finally dragged both feet to stop the bike. The back axle was smoking! I said to Elaine, "I think you saved our lives by leaning the right way." However, that was the last time I gave her a ride on my bike.

One Saturday morning, when I was feeding some young cattle, I noticed that one was missing. I walked along the fence and found that the wire had been cut and saw tire tracks. Pop called the sheriff. It took about twenty-four hours to determine that a group of young men in the area had a beef roast the previous evening. One of the boys was a neighbor, who had come home on furlough from the army. The sheriff asked Pop if he should arrest him. Pop answered, "No! We owe him more than one steer." The young man returned to serve his country the next day. Three weeks later we learned that he had been killed in Normandy.

The workload on the farm increased. Merle left for the army when I was a freshman. Wilbur married and left home about the same time. Paul joined the army when I was a senior.

I was able to continue playing basketball. We had a good team my senior year, even though we had no coach. I was captain and played center. I scored 230 points for the season, with a 14.4 point average per game. We tied with Stonycreek Township for first place in Class B, Somerset County. We played them for the championship but lost in overtime.

In the summer after my junior year in high school, I played baseball for the Rockwood men's team. I played third base and pitched. My best pitch was a "screwball" which broke in on a right-hand batter.

Route 653 Bridge – East of Rockwood, Constructed in 1913[9]

Shortly after it was built

View from bridge to the south

Looking up to the bridge from the railroad

I pitched only three good games. In one game, I hit four right-hand batters on the hands, with the "screwball." I had only two good games at the plate. In one game, the other pitcher was throwing a "knuckleball." Coach told me to keep my eye on the ball and not to swing hard. I had three singles. In another game the pitcher was throwing easy "curve" balls. I hit three home runs that game.

Jack Thompson went to Rockwood High. He lived 1½ miles from our farm. We rode our bikes together to school. He went into the Service at the beginning of his senior year. On graduation day, the principal gave Jack's diploma to me and asked if I would take it to his home. His grandmother cried when I gave it to her.

It was January 31, 1945. I was eighteen. My notice to be drafted was prompt and to the point. Who would take care of the two farms? E Jay was fifteen, Pop was sixty-six. It was about seven years since the war started. During those days, we stayed as close to the radio as we could. There was a lot of bad news, the worst of which was the number of casualties

However, I don't remember thinking much about it. I just got on the bus and went for my physical. It seemed to me like it took only a few minutes until they handed me my papers stamped 1A.

I walked a short distance and took notice of a sign on one of the tables—"Navy." I picked up some papers, answered a few questions, and signed my name. The officer at the desk remarked, "Congratulations, you're in the Navy now." I settled down in a chair to wait for the bus drive home. In a short moment my name was called for further examination. I received a new set of papers stamped 4F. I read the following on the report. "Systolic murmur at apex and along the left sternal lobe, enlarged heart in transfer direction." As far as I could tell, I was drafted into the Army and I enlisted in the Navy. Ten minutes later I was discharged. This was the shortest service record I've heard about: ten minutes in the Army, ten minutes in the Navy.

I went to my family doctor, Dr. Speicher, for a physical. He said, "I see no problem." Later a doctor at Penn State gave the same report.

We continued to make good use of the radio. If I came in late, my dad gave me a report on the war news. My dad liked other programs on the radio: *Amos and Andy, Fibber McGee and Molly,* boxing and country music.

Soon after Paul left for the Army, I quit baseball for the summer in order to carry a bigger load on the farm. Now it was just Pop, E. Jay and me on the two farms. Our cousin, Robert "Bobby" Ringer (sister Mary's son) and Lyle Walker helped some. Pop never drove the tractor or car, so I became the "key" machine operator. The biggest challenge was keeping the old equipment running.

After I graduated from high school, Pop told me that I could have all the milk money for the summer for college expenses. We only had eight cows that summer. My first cousin Mable Hay (Norman's mother) helped with the milking. I got up at 4:30 A.M. and drove to Merle's farm, milked the cows and filled the ten-gallon milk cans. We placed the cans of milk in the springhouse, so that the morning milk had as least one hour to cool in the cold water trough.

The milk truck came to pick up the milk every day at about 9:00 A.M. We milked by hand. Cousin Mable was very kind and I appreciated her help. After the morning milking was finished, I went back home for a big breakfast, which Mom always had waiting.

During this summer, Mom and E. Jay milked the cows at the home farm. Pop and E. Jay did the main home chores, which included taking care of the cattle, pigs and chickens. After I finished breakfast, we did field chores. We made hay, had threshing days, filled the granaries, as well as picking sweet corn for the cannery.

In September, when we were ready to fill the silo at the Bittner farm, I started cutting corn at noon the day before the crew came. Early the next morning, the twine tie-er quit working. After I had worked on it for about an hour, I looked up and there came three wagons ready to load. I called out to the crew on the first wagon that the corn cut at that point would be loose and harder to handle.

The last big job on the farm, as time was running out for me, was to cut and shock the fall corn. Merle was scheduled to come home in January 1946, after serving four years in India. I was leaving for college on October 15, 1945. Who was going to take care of the work at Merle's farm? I put an ad in the *Somerset Daily American.* We found a family who would live in the house on Merle's farm, take care of the cows and husk corn. We were fortunate and I was relieved.

In the meantime, I wasn't sure I had enough money for one year of college. Pop said to me, "Come, let's go see our senator." I followed him into the Senator's office in Somerset. Pop got to the point in a hurry. He said, "My boy, Floyd, wants to go to Penn State and we need some financial help." The Senator was quick to reply. He said, "I have a senatorial scholarship available. It's yours!" I was very thankful, for the Senator's generous offer and also for Pop taking the initiative to make this happen.

October 15, 1945, was the saddest day of my life. I had never been away from home, except for a very homesick week's visit at Luella's when I was twelve. I thought about Mom and Pop who would continue to work from dawn to dark. I thought about leaving the woods where I walked on Sunday and hunted in the fall. I thought about making maple syrup in the spring and cutting wood for firewood and posts. I would miss the big house, especially Mom's wonderful cooking. There were so many wonderful memories.

Now however, my work for the summer was over. I had reached a milestone in my life. I had graduated from high school. I now had a checkbook of my own with a balance of $601.35.

Chapter 14

College

Sketch by Carol Ogle

The day before I left for Penn State, Mom helped me pack, in one of Merle's old duffel bags, my meager amount of clothing, a tooth brush, razor, etc. I left the next morning at daylight. I walked down the same road, past Harrison Pritt's house, past the same rocks and fields, past the Long Farm. This time I carried with me everything I owned. I was ready to start my journey to State College by hitch-hiking. I wondered if anyone would give me a ride. No cars came by for some time, giving me time for reflection. I thought to myself, "Thanks, Mom! Thanks, Pop!"

I did get rides, and reached Somerset by 8:00 A.M. I took Route 219 to Route 22, 22 to Altoona, then Rt. 220 to State College.

I reached College Ave. by 11:30 that morning. My first order of business was to find a room. I entered the first building on campus and looked for a restroom. There I saw two boys with suitcases. I asked them if they knew how to find a place to live. "We have the same question," the bigger fellow replied. He said, "My name is Andy and I'm going to study Forestry." The other fellow introduced himself as Everett Corey and said he would be studying Chemical Engineering.

We found an office where a lady gave us a map and a list of about thirty people who rented rooms to students. We stopped for some hamburgers and began our search. At first it looked like all the rooms had been rented. Finally, at about 3:00 P.M. we found a single room and Andy took it. Everett and I had decided to room together since we would both be studying Chemical Engineering, so we continued to search for a place to stay. At about 6:30 P.M., I said to Everett, "Look, there's a cornfield, we're at the edge of town and we're running out of houses!" The lady in the next to last house told us, "I have one single left." We both chorused at the same time "We'll take it." This was the start of my college career.

As soon as she showed us the room, Everett headed for the bed and I took the couch. After resting a bit, we went back to Andy's and all went out for something to eat.

We stopped in front of a Greek restaurant. There was a sign in the window saying, "Help Wanted." We enjoyed a delicious meal and Everett and I had a job before we left. We later decided that restaurant served the best food in town. We worked there, waiting on tables, until we had our first exams. We both felt we needed more time to study, so we quit. I later found a job washing dishes in a fraternity house, with meals provided. I was able to eat and wash dishes almost as quickly as eating in a busy restaurant.

Everett dropped out of school after one semester.

I needed to find a place closer to campus. I found a place on West College Avenue. An elderly woman rented out rooms in her big house. She had three apartments on the second floor and three double rooms on the third floor. She gave me a job tending the furnace, for which I was thankful. I continued washing dishes for three years at the fraternity house.

My classes went well, except for my first English class, which I failed and had to repeat. My favorite class was Chemistry. There were many GIs enrolled at Penn State in 1945. There were about 300 students in the Freshman Chemistry class. We had three tests in Chemistry. The first one was a high school review. My grade was 74%. There were others in the class with a score of 100%. I got 84% on my second test. My final test score was 100%, the only 100% out of a class of 300. I also did well in a math class during the second semester with a grade of 100% on all five tests that were given.

I met Jim Rolls and he became my best friend at Penn State. He was also studying Chemical Engineering. Our last names started "Rol and Rom" and consequently we were always in the same classes and labs. My first year at Penn State seemed to go rapidly and summer break had arrived.

It was good to be home again. It was especially good to see Merle, whom I hadn't seen for four years. Merle suggested that I cut mining posts which were needed in the mines to prop up the roofs. So, I got the last order for mining posts for the Wilsoncreek Mine, as the mine closed soon after. I started working the day I arrived home. It took me several weeks to get back in shape physically.

I worked long hours, usually twelve hours a day. I used a one-man logging saw, an axe and a splitting wedge and a sledge hammer. I cut down trees, cut off the limbs, and cut the logs to the length needed. I split the big logs with the wedges and hammer. I used a small Allis-Chalmers tractor and pulled an old horse-drawn wagon. Following is the approximate count of posts delivered.

7000	32" Posts @ $.07 each
200	5' Posts
100	7' Posts
100	Hand-hewed ties for "dinkey" tracks

The total I received for the summer's work was $561.00. It was enough to pay for my Sophomore year.

During the following summer, I cut 12- to14-ft. beams to be used for other mine entrances. The money I earned was enough to pay for my Junior year.

The day after I came home from my freshman year in college, there was an incident involving Matt Kelly, a neighbor and frequent visitor to our farm. He and his brothers hunted groundhogs, picked apples off the ground and twice a year they came to get straw to re-stuff their mattresses. When Matt was an adult, he and his father worked in a coal mine in Blackfield. Matt's dad said that Matt worked only enough to buy whiskey and he had become an "alcoholic." When he drank too much, he would hallucinate and become hysterical, and see snakes.

That evening a car stopped at our farm. It was the Shroyer family, neighbors of the Kelly's. Mr. Shroyer came with the news that Matt had been shot. The Shroyer's had gone to Rockwood to get Dr. Saylor. Dr. Saylor told him to bring Matt to his office, but they were afraid to move him, so they came after Pop. Pop said to me, "Let's go!" I grabbed some blankets and old sheets from Mom and threw them into our old panel truck. Just as we were leaving, Merle came in for supper. I called out to Merle to come along, as we needed his help. Merle had been a medic in the Army. I jumped in the back and Merle drove.

We arrived at the Schaffer's house in about ten minuets. We found Matt wrapped in a blanket lying on the back porch. We could see a blood trail from the woods to the porch. As we loaded Matt into the truck, I could hear air passing into an apparent hole in his rib cage under his left arm. I sat on the floor in the back of the truck and held Matt in my arms. He was unconscious at first. Then he spoke. "Give me a smoke, Dave." Dave was his father. Later, he said, "Is that you, Floyd?" I had packed the torn pieces of bed sheets over the wound to stop the bleeding. Merle told me to hold his arm down to prevent more bleeding.

We arrived at Dr. Saylor's in about twenty minutes. When I told Dr. Saylor that Matt had a "sucking" wound in his rib cage under the left arm, Dr. Saylor said, "Don't bring him in here. Take him to the hospital in Somerset!"

It was another twenty minutes until we laid Matt on the table in the emergency room. A hole had been blown in the rib cage and there was another big hole in the back of his arm, exposing the bone.

They told us Matt was in severe shock and had nearly bled to death. He needed a transfusion badly, but he had a rare blood type. They took blood samples from his dad, Dave, and me, but there was no match. At 3:00 A.M., about ten hours after he had been shot, they found a match at the hospital in Mt. Pleasant. It was just in time to save his life.

We talked to his dad, Dave, after they had drawn our blood samples. He told me that Matt had been drinking and decided to end his life. He had gone to the Schaffer's to say goodbye to the girls, then went down the path into the woods. There he cut a forked stick and placed it in the ground and leaned his gun on the stick. He took a second stick to press the trigger.

As we stood in the hospital, Dave reached over and squeezed my shoulder and said, "Thanks for saving my boy's life."

Back at Penn State that fall, things were fairly routine with classes, studying, washing dishes and sleeping. I averaged from four to six hours of sleep at night and eighteen hours for everything else. I did go to all the football games and most of the basketball games.

I did not date at Penn State. I had met Jennie King, who lived in Somerset, right before my senior year in high school. We had one date before she left for her freshman year at Kentucky Christian College. We dated some the summer of 1945, before we both went off to college.

I missed classes two times at Penn State. The first time was at the start of my sophomore year. My eye had been bothering me and I went to see two different doctors. Both told me that I had some kind of growth in my eye. The third doctor finally diagnosed the problem. I had a small piece of corn stalk imbedded in my eye. He removed it and gave me a patch to wear over my eye for a short time. The last job of the summer before leaving for Penn State was helping Jim Pritts fill his silo. Apparently a fragment of corn stalk hit my eye while I was working at the top of the silo. It had been two weeks before it moved enough so that the doctor could see it. I had missed a week of classes.

Another time in 1946, I came down with the "mumps" and missed thirteen days. I decided to study calculus while I was in the infirmary. The day I went back to classes, an exam was scheduled. The professor told me that I needn't take it. I told him that I had studied and would like to take it anyway. The next day, as he returned the blue books, the professor announced "I guess I don't need to teach." I had made 100% on the exam!

By the summer of 1948, Jennie and I were married. I would need to find a suitable place for us to live and there didn't seem to be anything available. My sister Mary's husband, Bob Ringer, had a trailer for sale. I was able to borrow $750.00 from Wilbur and bought the trailer. I found a man who agreed to move the trailer to State College for us. It was set up at Spring Valley Trailer Park, which was about three miles from town. I was back to walking to classes for a short while. Then Wilbur and Merle sent a bicycle to me which was a tremendous help. At this time, Jennie stayed home.

That fall, Mom and Jennie's mother came to visit us. While they were there, we received some sad news. Jennie's father had been killed in a farm accident!

After nine months of living in that trailer, I graduated from Penn State with a B.S. in Chemical Engineering and a 3.3 GPA. We put up a sign, "For Sale $250.00" on the door to the trailer, our first "home," and came home for the summer. We stayed at Jennie's mother's home in Somerset. I worked that summer to do some clean-up chores on Jennie's father's farm. In August 1949, our first child Sylvia was born. We brought her home from the hospital when she was three days old and immediately realized that something was wrong. Dr. Saylor had her admitted to the hospital in Johnstown. Jennie, in the meantime, was bedfast from some instrument damage during Sylvia's birth. I drove down to the hospital to be with Sylvia. I can remember praying and asking for God's intervention. I went so far as to try to bargain with him. Sylvia lived only thirteen days.

Jobs were scarce in 1949, so I applied for graduate school at Penn State. I was accepted, but later told that they could accept only one student from our class. The Teaching Fellowship went to Ted Williams, who had the best grade point average in our class. I needed to make other plans.

I drove to Lewisburg, Pennsylvania, home of Bucknell University. There I met with Dr. Cooper, head of the Chemical Engineering Department. He offered me a University Scholarship, which I accepted in place of a Teaching Fellowship. It would pay all of my tuition and the cost of books. Along with the money I had earned that summer and some money we received from Jennie's father's estate, we were able to buy a car and pay our living expenses. Jennie found a job assembling electrical wiring for auto motors.

I met Dick Fulmer at Bucknell. He became my closest friend. During Christmas vacation 1949, he came home to the farm with us. This was a new experience for him. He shot his first rabbit at the Long Farm. There was about a foot of snow on the ground.

After receiving my Master's Degree in nine months with a 3.9 GPA, I decided to go on to get a Doctorate. One of the professors at Bucknell suggested that I go to the University of Cincinnati to work on my Doctorate. We drove to Cincinnati and I registered for fall classes. I applied and received a Research Fellowship, for which I would also receive a salary. In the fall of 1950, Jennie and I were ready to go to our next "home." On the way from Bucknell, we stopped at my parent's farm for two days. When we left the farm, our car was loaded with three bushels of sweet corn and twenty quarts of canned yellow cherries. Thanks, Mom! We lived in a one-room apartment for three years. Jennie worked as a secretary at a leather goods company during our stay in Cincinnati.

At the University of Cincinnati, I worked from 8:00 A.M. until 5:00 P.M., except when I had classes and also on Saturdays from 8:00 until 12:00. I worked on three projects during the three years at U.C. They were:

1) Adhesion of rubber to steel for shock absorbers,
2) Decontamination of mustard gas, and
3) Preparation of rocket fuel.

I wrote my thesis on oxide replacement for hydrogen peroxide for rocket fuel.

In the lab we had a number of solvent fires. Usually the fire started with an explosion of petroleum fractions. The most serious fire blew onto one of the students. Everyone ran to help. I was the third person there. The two students ahead of me ran through the lab doorway with a fire extinguisher in hand. Don, the student, ran out the door with flames extending from his belt to the ceiling. I grabbed him and wrapped a fire blanket around him and snuffed out the flames, while students with the fire extinguisher put out the main fire. The gasoline had covered his throat and chest, also leaving him with burns on his face. Those on his throat were third-degree burns. He was fortunate not to have burned his lungs.

I walked back to my laboratory after they took Don to the hospital. Dr. Cunningham, my research instructor, was sitting in his chair, pale and shaking. He said, "You know, I think you saved his life."

I received a PhD in Chemistry in 1953 with a 3.65 GPA.

While at U.C. we were able to save enough money to pay Wilbur the $750 we had borrowed to buy the trailer. We also had enough to buy kitchen and living room furniture, when we made our next move to Midland, MI. Three days after graduation, June 1, 1953, we were living in a duplex in Midland, Michigan. I started work the next day at Dow Chemical. After eight years at Penn State, Bucknell and the University of Cincinnati, I had finished my education and begun my career as a Chemical Engineer!

Chapter 15

Where Did They Go?

Photo by Carol Ogle

Ephraim and Mayme (Pop and Mom): My father was born in 1878 in a small log cabin located just across the road from where Uncle Milt's house was built later. Uncle Milt was also born in this cabin in 1881. The family was living in the frame house when my grandfather Levi died in 1889. Sometime between 1881 and 1889, Levi moved his family from the log cabin to the frame house. This was Pop's home until he died in 1964. Pop spent almost all of his life farming and working in his woods, except for a short time when he worked for the electric company. He served on the school board and was a constable and Fire Warden for Black Township.

Beverly and I came home for Pop and Mom's fiftieth wedding anniversary in 1963. Pop's last words to me were "Take care of Mom."

Pop died March 25, 1964, five months before my daughter Cindy was born. My sister Helen, who was a nurse, was at Pop's side when he died.

Luella, Betty and I were at the funeral parlor the day before the burial date. Pop often said to his children, "Give me my flowers now while I still live, and when I die, put my Bible with one of the little red roses in my hand. Luella said she would like to look at his Bible. We decided to borrow it and take it out of his hand one last time. As Luella gently turned the pages, I thought about Pop's stories about Grandmother Mary Ann reading the Bible when Grandfather Levi died and how Pop started to read the Bible when my Grandmother died. At that time he could barely read. He pointed to each word and read in a loud whisper—and often asked Mom how to pronounce words. He read every moment he could. Mom said he was a pest. He would come in the house for short periods of time with mud on his boots or rain water dripping from his clothing. He sat in his favorite chair next to the stove. Each time he read the Bible from the beginning to end he put a red mark on the inside of the book cover.

There were thirty-two red marks when he died.

Mom came to work for Grandmother Mary Ann about 1912. She had been previously married to George Bockes. George was killed

Ephraim and Mayme in 1960

while working on the railroad. The workers had laid planks across the track as a temporary means to drive a team of horses and wagon across the track. When a train suddenly showed up, George rushed to remove the plank. The train struck a plank which struck George and caused his immediate death. They had an infant daughter, Mary. Mary was raised by George's parents. Mom and Pop were married in 1913.

Mom always wrote letters to all her children, as each left home. She kept everyone informed of any news. We always looked forward to her monthly newsletter.

Just after Pop died, Mom became seriously ill. In spite of a lot of pain, she wanted to visit her family as much as she could. Luella brought her to Harborcreek; then we brought her to Midland, Michigan for a visit. She was very uncomfortable riding in the car. We had our doctor see her as soon as we arrived. He said she had terminal cancer. Luella's daughter Peggy had taken Mom to her doctor who made the same diagnosis. In spite of being very ill, Mom enjoyed visiting and seeing our two daughters, Beverly and Cindy. She also enjoyed seeing our new house which was near completion. We took Mom home for the last time. Mom died October 25, 1965. Helen was with her when she died.

Mom was a kind, generous, hard-working, loving mother.

Pop and Mom are buried in the Hauger Cemetery along with Mary Ann and Levi, Polly and Solomon, and Uncle Alex.

Hauger Cemetery

The Hauger Cemetery is along Route 653 (the road to Garrett) at the intersection with the road to Wilsoncreek. A small burial ground started in 1859 when members of the Simon Hauger, Sr. family donated land for a burial ground. The Church of the Brethren stood from 1885 until 1937 in front of the old section of the cemetery.

Photo by Clarke Romesberg

#1 Mary: Mary lived with her grandparents. After high school, she married Robert Ringer. Robert was a truck driver and a painter. They had three children, Jean (adopted), Robert (Bobby), and Floyd Wilbur. Our family enjoyed their visits. Robert was a good ball player. He took us (Merle, Wilbur, Paul and me) to two Pirate ball games. Paul and I rode in the rumble seat. I can still recite the 1937 Pirate lineup. During the war, Bobby helped us with the farm work. Bobby and E. Jay became close friends. They enlisted in the Navy together after High School. Mary died of colon cancer on April 25, 1986. She was 73.

#2 Luella: Luella was the first sibling to leave the farm. In 1932 she married Woodrow Romesberg. They had two children, Peggy and JoAnn. They lived for a short while in the little house along Route 653, near the Blackfield dinkey track. (Solomon and Polly had lived in this little house before they moved to the farm.) Then Luella and Woodrow moved to a bigger house near the Matt Romesberg residence.

I was awakened late one night when I heard Mom crying. Merle and Wilbur had just come home from Luella's house. Woodrow had been ill for several days and suddenly became very ill. Mom and Pop dressed and walked the dark path through the wood to Luella's house. The doctor did not know the nature of Woodrow's illness. He died at daybreak.

Mom and Pop came home early in the morning. Mom sent me to the hen house for eggs for breakfast. I remember thinking that this was a sad day and that I should remember that Wood died when I was nine years old.

About a week after Wood died, his mother came to visit Mom. She told Mom that she had put Wood's shoes away. She said they still had mud on the soles from the last time he walked home. Then she cried. Her son had died at a very young age.

For the first eight years after Luella left the farm she lived close to the farm. I could hardly wait for Sunday to come. She brought her family to dinner nearly every Sunday.

Luella was married the second time to John Ogle, in 1937. They moved to Boswell where John worked in the coal mines and had one child, John William. When John left the coal mines, they moved to Erie, Pennsylvania.

John Ogle grew up in Stoystown, PA where he attended high school and got letters in football and basketball. He left high school before graduation to work in the coal mines. John would later tell about lying on his back on the floor, deep in the mine, picking coal from the seam above him. One day some of the ceiling collapsed and one of the miners was killed. John said the crew leader had the body moved to the elevator shaft and ordered that it would be taken to the surface at shift change. John said that when he worked in the mines they would ride the elevator down in the early morning before sun-up, and come back up after sunset, working six days a week. He would see the sun only on Sundays.

John liked to visit the farm. He often took Pop and me fishing. One day he said he was getting out of coal mining because of the dust. He said he was collecting tools to become a carpenter. Wilbur gave him a chisel. John sharpened it and later gave it back. He moved to Erie and got a job at the Hammermill Paper Company. He got a second job working the "Victory Shift" at the Erie Forge during the war. Later John and Luella moved to Harborcreek, PA, just east of Erie, where he and his brother David raised mink. Luella and John continued visiting the farm on weekends. Mom always had fresh home made bread, just out of the oven. John especially liked apple butter on fresh bread with milk. Eph and Mayme often visited John and Luella at Harborcreek during the summer months. Sometimes during those visits, John would take Eph with his fishing pole and a sack lunch to the pier at Lake Erie on John's way to work. Eph would fish from the pier all day and until John picked him up after work.

John took up carving as a hobby in his retirement years. He made beautiful carvings of birds and animals. John died in1988 after a long illness with cancer.

While living at Harborcreek, Luella was an elder in the Harborcreek Presbyterian Church. She was also very active in cub scouts, serving as a den mother for several years. She worked in the high school cafeteria, earning money to send John W. to Penn State.

Peggy Jean was born 23 January 1933. She married James Howard Nelson in Harborcreek, PA. Peggy was a graduate of the Hamot Hospital School of Nursing and is a retired R.N. Jim is a retired safety inspector for Penelec. Jim and Peggy owned a camping equipment store for fifteen years and then an antique store for three years. Finally they are fully retired. Jim and Peggy had three children;

1) Richard Alan, born 1954. Rich married and moved to Vermont, divorced and now works for Smugglers Notch, a ski resort. He has eight cats at the present time.

2) Sueann, born 1956. Sue married Arthur Mills who worked for the State Department and they were posted all over the world. Art is now retired from the Foreign Service and is teaching at a local college. Sue recently graduated from the Culinary School at Mercyhurst. They are settled in North East, PA with three children—Katherine Elaine is now attending Penn State, Andrew James, a new high school graduate, will also attend Penn State in the fall of 2008, and Edward Thomas Earl, a junior in high school and a competitive diver, will go to Mercyhurst after graduation with hopes of becoming an operating room nurse. When Sue and Art married, Art had

a little girl, Mollie, from his first marriage. Mollie married Jesse Mann and Eliza was born, making Peggy and Jim great grandparents.

Five generations of the Ogle family, with the newest member, seven-month Eliza Mann. Seated at right is Eliza's great-great-grandmother Luella (Romesberg) Ogle. Behind Luella is her daughter, Peggy Nelson, and next to her is Luella's grand-daughter Sue Mills. Luella's great-grand daughter Mollie Mills Mann is holding the great-great-grand-daughter.

3) Bonnie Lou, born 1960. Bonnie married Don Duckett and they reside in North East, PA. Don works for Welch Foods and Bonnie is a therapist assistant at Health South, a Rehab Hospital. They raised two girls, Beth Shannon, now an RNBS working at Health South, and Holly Sue, attending college in Boston and working on her Masters in Psychology.

JoAnn married John (Jack) Caruso. Jack lived with his mother. When Jack was three years old, his father divorced his mother, re-married and then moved to Florida. JoAnn and Jack had six children: John Michael, Kathleen Jo, Steven Robert (died as an infant), Debora Ann, Terri Lynne, and Laurie Ellen.

Jack and JoAnn left Erie in 1960 for Jacksonville, FL where Jack's father had promised him a job with his electric company called John's Electric. Jack and JoAnn loaded their belongings into a small trailer and the family into their car and headed south. After a short distance, they had a flat tire. This happened three times. Apparently old tires did not hold up on a trailer with no springs. With the delays, Jack called his dad to tell him they would be a day late. After a long pause, his dad told Jack, "It doesn't matter. We don't want to hire you." Jack hung up the phone, paused and cried. They had no choice but to keep going. They arrived at Jack's dad's home and to their surprise, Jack's dad had persuaded his wife to hire Jack.

Jack got a second job tending a gas station. At Christmas, his mother-in-law gave him a present—a ball-point pen. Jack couldn't resist. He quit. Jack then got a job selling cars. There were two other salesmen. The first thing Jack did was to turn on the lights and work until 11:00 P.M. At the end of the first month he sold more cars than the other two together. His boss made him manager of used car sales The next month Jack was promoted to new car sales.

Jack was not only an expert at selling cars; he knew how to manage money. When Jack's boss was ready to retire, Jack made a deal with him to buy his business. Within ten years, Jack would reach a lifetime goal: his Chrysler-Plymouth dealership would become number one in the country along with his new Jack Caruso's Regency Dodge. The success of the two businesses have allowed JoAnn and Jack to travel with the jet-set, staying at the best hotels and dining in the best-of-the-best restaurants in major U.S. and international cities. They have flown on the Concord from NYC to London and returned to NYC on the Queen Mary 2; rode the Bullet train in Japan and the Eurorail in Paris to Switzerland; also the Chunnel train under the English Channel. They have walked on the great wall of China and crawled through the pyramids in Egypt. Toured the Kremlin and walked in Red Square; sat in the coliseum in Rome and viewed the Sistine chapel before, during and after its restoration by Fuji. They danced in the Eiffel tower in Paris with the Mayor of Momart and slept in Ashford Castle in Galway, Ireland. Toured castles in Germany, Spain, England, Scotland and the churches and museums of the world's greatest artists. The Hermitage, the Louvre, the Prado, Versailles, Monaco, Vatican and the temples in Hong Kong, Bangkok, Shanghai and Beijing. They viewed the majestic Mt. Fuji and Denali and the Swiss, French, Italian and German Alps. But their biggest adventure was getting to visit Antarctica. They boarded the *World Explorer,* a German ice cutter ship, at Punte Arenas, Chile, and sailed across the straits of Magellan around Cape Horn and on down to visit nine science stations in Antarctica and to sit with the penguins on ice glaciers and look a Menke whale eyeball to eyeball while sitting in a zodiac (rubber raft). The trip back to Ushia, Argentina brought twenty-seven-foot waves and tossed everything and everyone around for several hours, but they were wearing the sea bands on their wrists and it kept them from getting sea sick. Most on the ship were not that lucky. Jack and JoAnn are currently living in a $3 million condo on the Atlantic Ocean at Jacksonville Beach, Florida and are quite content to stay at home.

JoAnn had come a long, long way from the little eight-month-old baby girl I saw my dad carrying as he led Peggy by her hand. It was the day after her dad's funeral. We were walking down the dusty dirt road leading from Matt Romesberg's home to Luella's weathered house located among brush and weeds near the deserted town of Blackfield. But JoAnn will be quick to tell you she still loves to visit the homestead and she still has the fondest of memories of all her aunts and uncles and of Gramma and Grandpa and that wonderful farm.

John William was born 5 November, 1939 in Boswell, PA. My sister Luella with her husband John Ogle and John W. came to visit us at Midland. They went on the Dow plant tour for visitors. John W., then 14, was impressed with the processes, the products and, especially what engineers did on their jobs. When they completed the tour, John W. remarked to his parents that he was going to Penn State, get a degree in Chemical Engineering, and then come back to work for Dow Chemical. Ten years later he was back with his Master's Degree in Chemical Engineering. While at Dow, John's work included plant design and the manufacture of various plastics and hydrocarbons. He worked in Midland, Michigan; Plaquemine, Louisiana; and Lake Jackson, Texas, as well as temporary startup assignments in Altona, Australia and at Terneuzen, The Netherlands. After retirement from Dow, he continued to work in his own firm as an engineering consultant to chemical manufacturing operations, primarily on environmental issues. While at Penn State, on August 20, 1960, John married his high school sweetheart Carol Seyschab (b.12/24/39), who earned a degree in Art Education at Penn State. They had three sons:

1) John William, Jr, born 5/10/1961 in Bellefonte, PA. After graduating from high school in Lake Jackson, TX, John, Jr. attended one year of college in Texas and then worked in a local cabinet shop. He married Janet Edwards and one day packed all his belongings into a small Toyota pickup truck and, accompanied only by his dog, drove to Alaska to take up work as a cabinet maker. Janet later followed him to Alaska and they had two children, Dylan Ross (b. 3/3/86), a high school and college soccer player, now attending school in Idaho, and Alyson McKay (b. 9/13/87), who worked for BP while attending high school in Anchorage, now attending Texas A&M in pre-med. John and Janet divorced and John moved to Homer, Alaska to take up commercial fishing and crabbing. There he met Denise Stockton, who was visiting Homer on summer break from the University of Kansas. Denise bought crabs from John at the Homer boat dock and resold them to tourists. That started their relationship and they married and had three children, Corinne Kemsley (b.10/31/91), Sophie Rebecca (b. 1/8/94) and John Leonard (10/15/98). John got his Coast Guard license and he and Denise operated a sports fishing business at Anchor Point, Alaska for several years. John is currently operating an Alaskan tourist business (Bear-Paw Adventure) and is constructing high quality log buildings using local spruce trees. Denise worked for some time at the local Russian school, and works now at Homer High School.

2) Jeffrey Scott, born 12/25/63 in Midland, MI. After high school in Lake Jackson, Texas, Jeff attended The University of Texas where he earned a BA in data processing. He married his high school girlfriend Donna Jones, who also graduated from UT with a degree in accounting. Jeff and Donna originally settled in Pearland, Texas, where Donna drove north each day to Houston to work at an accounting firm, while

Jeff drove south to Freeport, Texas to work at Dow Chemical. Later they moved to Lake Jackson where they both currently work for Dow Chemical Co. Jeff does taxidermy work as a hobby, and his products include some of the big game animals taken on Alaska hunts with his dad and John Jr., including moose, caribou, Dahl sheep, bear and other critters as well as wolverines, bobcats, ducks and Texas white tail deer. Jeff and Donna had two children, Alexandra Karyn (b. 4/28/92) who is very active in sports, especially softball, and Hunter Scott (b.10/2/95), who is also active in sports including football and who also loves to hunt and fish.

3) James Richard, born 5/23/67 in Midland, MI. After high school in Lake Jackson, Texas, Jim earned a BS degree in Mechanical Engineering at The University of Texas. While at UT and on August 13, 1988, Jim married his high school girlfriend, Michelle Moore. They had six children, Krystal Nicole (b. 8/20/89, who is currently in pre-med at UT, Lauren Claire (b. 11/6/91), twins Heather Elise and James Ryan (b. 5/30/94), Jakob Ryker (b. 6/18/03), and Jaeger Roman (b. 3/22/05). After graduation from UT, Jim went to work for Dow Chemical and lived in Lake Jackson, Texas for seven years and then moved to Dripping Springs, Texas, which is near Austin. He is currently located at Midland, MI, where he is working on a new industrial plant that will produce a product intended to reduce particulate emissions from diesel-burning engines. Michelle is taking time out from her nursing education to raise the children.

#3 Merle: Merle was the oldest boy in our family. He was a skilled workman. At seventeen he became a taxidermist. He mounted deer heads, small animals and birds. He was a good mechanic, skilled at repairing equipment. He was a mason and a carpenter. At age eighteen he built our stone spring house with an underground stone and cement wall to retain the water.

Merle had a serious accident when he was eighteen. He was leading a horse with a chain. The horse bolted and the hook on the chain mangled his middle finger on his right hand. He lost one-half of the finger.

Merle was a supply sergeant in the Army Medical Corps during WW II. He served four years in India.

When he retired from the Army, he married Dorthy Schaffer, daughter of Walter and Cora Schaffer. They had three children, Nancy, Merle, Jr. and Larry.

Merle took over the Bittner farm in 1946 when he came home from the service. They lived in the big house on the home farm, while Merle remodeled the home on the Bittner farm. Over a period of time, he remodeled the barn, built a big machine shed, two silos, and developed a dairy farm. Over the next twenty-nine years, Merle missed only two milkings.

While Merle and Dorthy still lived at our home at Wilsoncreek, Nancy was born. She was the last family member born in the big house. When Dorthy went into labor early, Dr. Musser told Merle to come to Somerset for some medication. Merle asked me to go along. When we retuned and opened the door, we heard a baby cry. Nancy came early!

Shortly after Merle and Dorthy moved into the old house on the Bittner farm, Merle built a new house. After they moved to the new house, Merle bought the Keeler farm. My friend Bill Shaw lived on this farm with his grandmother, Mrs. Keeler.

Merle cleared fence rows and some wooded areas to improve the Keeler farm. On one side of the property there was a small wooded area (about three acres) with an extensive outcrop of rocks, which made it very appealing. At the very edge of this plot there once was a little church. We went there for summer picnics. I liked walking through the woods to look at the rocks.

Soon after Merle bought the Keeler farm, he built a house on the small wooded lot opposite to where the church stood. Merle and Dorthy moved into this house and Merle, Jr. and Hazel, who first lived in a trailer house, moved into the farm house and took over the operation of the farm. Then Merle built a third house for Nancy and Harry on the spot where the little church stood.

Nancy married Harry Tunstall in 1965. They had two children.

Merle II served two years in the U.S. Army. He was stationed in Hanov, Germany. He married Hazel Beck. They had one son, Merle III. Merle II helped Merle in the operation of the farm. Hazel taught school. Merle III graduated from the University of Pittsburgh with a PhD in Physics. His work has involved development of equipment to improve radiation treatment of cancer.

Larry, while in high school, designed and built a laboratory-size unit to make methane from cow manure. He demonstrated his unit by opening a valve and igniting the effluent gas with a match. He later went to college and studied mechanical engineering.

#4 Wilbur: Helen and Doris were best friends in high school. Helen introduced Doris to my brother Wilbur. They married in1942. They operated the dairy farm established by Doris' parents, Simon and Susan Enos. Wilbur had four children.

1) *Sue Helen* – teacher, named after aunt and grandmother
2) *Shirley Joan* – teacher
3) *Wilbur George*
4) *Wayne Enos*

Wilbur George died from gasoline fumes while attempting to clean tar from a 55-gallon drum. He was twelve years old. After Doris died from cancer in 1964, Wilbur sold the farm and moved to Rockwood. Wilbur died at eighty-one in 1996.

Following are comments I made at the Eph and Mayme family reunion in 2006.

1) Luella claims Wilbur was the most mischievous sibling in our family. He once threw an old clock up in the air and then watched it fall and hit him on the head. Another time, Mom had to call Dr. Saylor when Wilbur had stuffed sand in his ears. In another case, a serious accident occurred when he was

cracking nuts with a hatchet. Helen reached for a nut just at the wrong time, and got her finger cut off!

2) When E. Jay and I were quite small, we played games on the floor. We both had a similar experience. We always played until we fell asleep. Then we could not remember how we got to bed. One time E. Jay fell asleep and I just lay there looking at him. Suddenly, a pair of hands reached down and picked up E. Jay and carried him to bed. Wilbur was very considerate of others. He never went to bed until he knew E. Jay and I were sound asleep.

3) One summer day in 1932, I was with Wilbur while he was cutting brush in the fencerow along the road just past Uncle Milt's farm. He stopped cutting with the scythe and said, "I don't want to cut this beautiful red oak tree." It was about six feet tall. I said, "Don't cut it." Then Wilbur said, "I'll let it grow for our little brother E. Jay." We often called it "E. Jay's tree." Every year we watched it grow, then some time after we left the farm, the coal strippers uprooted the tree with a big bulldozer. The tree is gone now and so is Wilbur. The memory of Wilbur's care for a beautiful tree still lingers in our minds. The loss of this tree left a hole in the landscape and the loss of Wilbur left a hole in our lives.

4) On Saturdays, during the winter months, we usually worked in the woods. It was fun when it snowed. One Saturday we had a heavy snowfall while eating lunch. After lunch, Pop left first and was well ahead of us four boys. We walked in his tracks in the fresh layer of snow. There were Merle, Wilbur, Paul and me. Four inches of heavy snow covered the ground and all the limbs on the trees, stumps and small trees. I looked past the tree in front of me and I could see how exacting we were in following in Pop's tracks, step by step. The scene was beautiful. We boys were like a string of four black dots moving across a huge blanket of snow. For each step I had to jump a little to reach the span between each step set by my

father. Suddenly the little caravan of dots stopped. The snow-covered forest was silent. Wilbur held onto Merle's shoulder. "Look," Wilbur said. "Look at the path we're following. Will we be able to make a path that straight and narrow when we're as old as Pop?"

5) One evening, just after dark, Wilbur walked into the house. He had just come from our neighbor's house, Harrison and Mamie Pritts. He walked to the opening into the dining room, leaned against the door jam, took off his hat and rubbed his head. I was sitting just to the right of the chimney where I could feel the warm air and the warm plaster on the brick. Mom looked at Wilbur and then asked, "How is Mamie doing?" Wilbur looked down as if not wanting to answer. Wilbur looked up and replied, "Mamie is dying. I'm going back. I promised Harrison I would sit with them tonight." Years later I thought of how considerate that was, coming from a fifteen-year-old boy.

6) At our Romesberg family reunion in Rockwood, PA, in 2006, I was mindful that Wilbur's three children (Sue, Shirley and Wayne) were present. I had a short and final message especially for them. When I came home from Penn State during school breaks, I usually saw Wilbur often, sometimes just long enough to say hello. Wilbur would usually ask the same question—"Hello, how are you doing? Are you getting enough to eat?" Then he would reach in his pocket and pull out a $20 bill and say, "Here, this might help you a little." I stopped and paused for a long moment. Then I called out, "Wayne, come up here and help me finish this." As he arrived, I reached in my pocket and pulled out a $20 bill and handed it to Wayne. He drew back. Then I said, "Take it, it's just one of the twenties your dad gave to me. It is a reminder of all the good things your dad did for me."

#5 Della: Della was the fifth child. She had eczema and many allergies. Mom did all she could to doctor her. She made arm and leg socks so Della wouldn't scratch herself so much. Della got pneumonia and died at age five in the hospital in Johnstown. Pop said it was a sad day when they carried the little casket into the living room.

#6 **Helen:** Helen was the first in our family to go beyond high school. She enrolled in nursing school in Washington, DC. One day when she was about halfway through the course, she retuned home and said school was too hard and that she wanted to come home. I remember Pop talking to her alone in the house. I don't know what he told her, but she went back to school and that was the last we heard of school being "too hard." It wasn't long until she was a registered nurse working in a hospital in Washington, DC.

Helen married John Forsythe and had two daughters, Judy and Patricia. John died seven days after he had an automobile accident on the way home from bowling. Apparently he lost control when a dog ran in front of him. The car struck a telephone pole and John was thrown from the car. His head struck on the shoulder of the road and he never regained consciousness. Just before he died, he opened his eyes and said, "Who will take care of Patricia?"

Helen later moved to Somerset, PA. She worked at the state hospital.

While in Somerset, she opened a childcare center, some time around 1960. The venture was unsuccessful. Not enough people needed such a care center in 1960.

Helen moved with Judy and Patricia to Washington State. Patricia was a slow learner and was an extra burden for Helen. Patricia died when she was forty-nine. Helen later returned to Somerset. She died when she was eighty-one.

#7 **Paul:** Paul was born on the farm in 1922. He was the tallest and strongest of the five boys. Sometimes, while still a young boy, he would disappear. Generally we found him in the barn in the hay-mow fast asleep. Paul was a hard worker. He cut (with the sickle) and shocked more corn than anyone I knew. He was the one that was assigned to the heavy jobs.

Paul served in the U.S. Army from 1945 to 1948. He married Violet Schaffer, sister of Merle's wife, Dorthy. They had seven children.

*Paul and Floyd
in 1933*

He and Violet lived on the home farm where they raised their children, James, John, Joseph, Jeffrey, Jane, Joy and Janice.

Paul operated the home farm for eighteen years. With Merle's help, he built a new barn and silo. However, Paul's heart wasn't dedicated to farming. He wanted to follow the path made by his grandmother, Mary Ann (Livingston) Romesberg, and his dad, Ephraim Romesberg. That path was to read the Bible and preach the gospel. He left the farm to become a preacher. Paul retired from the ministry and they moved to North Carolina.

Paul's Career in the Ministry
(written by Paul on August 19, 2008)

On July 9, 1961, after repenting of my sins, I was baptized in the name of Jesus Christ for the remission of my sins, filled with the Holy Ghost, speaking in tongues. St. John 3-3,4,5, Acts 2:38. In 1963 I was called of God to preach the Gospel of Jesus Christ.

In 1965 I was ordained into the ministry by the Bishop Board of the Church of Jesus Christ, Somerset, Pennsylvania.

I helped build the Church in Somerset, PA from the foundation to finish. I spent around twenty years there. In 1968, I moved to Oxford,

PA and pastored the church there for three years. The Pastor had both legs amputated above the knees. I stayed there until he passed away. Then I moved to Rockwood, PA.

I evangelized out of the Church in Somerset and in many assemblies in many states. Here is a list of the states. Pennsylvania – 22, Maryland, Delaware, Virginia, West Virginia, Ohio, Kentucky, Tennessee, Michigan, California, Indiana, South Carolina, North Carolina, Georgia, Florida, etc. In 1983, I volunteered to go with a missionary to the Philippines. I preached my way from Pennsylvania, through Kentucky to Tennessee. We flew from Nashville, Tennessee to Fort Worth, Dallas, Texas to San Francisco, California. I preached two weeks in California before leaving San Francisco, going down through San Fernando Valley close to Los Angeles, in Bakersfield. There we had the privilege to travel across the most crooked street in the world, which is in San Francisco. From San Francisco we boarded a 747 Pan Am flight, with 500 aboard, plus all the luggage and tons of fuel. We landed in Hawaii and Guam before landing in Manila, Philippines and then flew to Cebu Island in the Philippines. We traveled mostly by boat from island to island. Some of the boats were three stories and some were small boats with rudders on the side, so they wouldn't turn over. We traveled to around a dozen islands. I had the privilege of baptizing several Philippine people in the Pacific area. What a beautiful scene it was. The water is clear blue. I spent six weeks in the Philippines and three months in all, till I returned home. I had a wonderful experience.

In 1984, I started a Church Home Missions work in Champion, PA. I held services in a store front. After six months, I bought property in Indian Head, PA. Here we remodeled a barn, a large trailer and a two-car garage. Now we had ample space for Sunday school and church services.

James, John and Joseph worked for DuPont in Wilmington, Delaware. After twenty-eight years, James retired and now operates a consulting business. Jeffrey was a carpenter for many years and then became a salesman. Janice married a minister and is an assistant chiropractor. Joy and her husband operate a home-care and antique business.

#8 Betty: (written by Betty Romesberg Clarke) I was born, in a lonely, quiet place: devoid of luxury and comfort, but in a Godly atmosphere. It was here, at home where all ten children first saw the light of day. Sometimes I was embarrassed to belong to a poor family and longed to see the outside world. But there were many compensations: like all the wonderful food prepared by our mother; the times to dream silently while laboring in the fields, milking the cows and doing all the other various farm chores; the benefits of much walking to be appreciated later; the walks through the woods to the general store in Wilsoncreek and to the distant parts of the big farm; the daily walks to grade school and many times to Rockwood through all kinds of weather.

After finishing school I had an opportunity to go to West Palm Beach, Florida. That was like going to the moon. Thus having left the farm, I was ready to seek my fortune. I began in Erie, PA (thanks to a wonderful sister), where I began a ten-dollar-a-week job as a student stenographer. When there was nothing more to learn, I did all kinds of work for a credit reporting bureau. Then I took a job as a dental assistant. This led to Temple University in Philadelphia where I became a registered dental hygienist. I felt that I had conquered the world as I entered an office in a well-known practice in Wilmington, Delaware. After joining the Ski Club, I met my future husband, a PhD chemist with DuPont de Nemours. After marriage, we moved to Massachusetts and Connecticut and enlarged our family with a girl and a boy, Tiffany and Kevin. Then we moved to Switzerland and stayed. Our daughter decided to go to medical school at Georgetown in Washington where she met and married another doctor. Our computer consultant son married a French girl and lives in France with their two children. We now have two U.S. grandchildren and two here in Europe. Life here is rewarding and beautiful. Restaurants are inviting, relaxing and the food wonderful. Traveling is equally interesting and we have skied most of the well known resorts, plus travels to almost every country in Europe and the Middle East. Now we relax part-time in our mountain home where pollution is still absent and quiet reigns. We walk, ski, and relax and enjoy nature and its tranquility. Many friends and relatives come to enjoy this paradise and my husband often says: "Aren't you glad you worked hard when you were young?"

I am grateful for the family and events that shaped my destiny and gave me a most wonderful life.

Honor to my Husband
by Betty Romesberg Clarke

When Ken knew that time was fleeting,
He asked me to tell of our meeting
The 13th of November 53 a beautiful day,
On a ski piste in the USA.
We were there with others from the YMCA
To clear the piste for the season not far away.
Donnerkopf was very small
But better than no piste at all.
When I got home my phone was ringing.
I heard the voice that started my heart singing.
He came to call every day
And said he wanted to keep the other guys away.
So on April 9, 1954
We were married and went to the Jersey shore.
For a long weekend we did hope,
Until he stepped on his glasses and they broke.
The next week and a big surprise
At his boss's house and before my eyes
A table with bridge cards and beer.
I was captured that was clear.
I did not know his passion to play
So the lessons began that very day.
His patience to teach,
Like his love of life—never did cease.
At Arthur Murrays we learned to dance,
The Caper Club gave us the chance.
Its various themes and floors so spacious
Kept us trim in the right places.
We moved many times interestingly so

And went wherever Dupont told us to go.
We loved life in the spooky old gate house,
Ancient and High,
One floor atop another,
It reached to the sky.
The old picket fence around the yard,
Stands there still,
A soldier on guard.
We worked in the garden
Where wild flowers grew.
It was a bit of Heaven we knew.
We toiled, mowed and weeded
Much time and care was needed.
The robins and squirrels all around
Kept a watchful eye
For food on the ground.
Then the time came to say goodbye.
We looked at the old walls.
They seemed to cry,
Don't Go! There's more to do here.
But we had to journey on
Both far and near.

Betty

Photograph courtesy of
Betty Romesberg Clarke

Kenneth G. Clarke
1927 - 2008

OUR CHURCH WINDOW

He makes wine to gladden the heart
and bread to sustain us.

Psalm 104:15

Given in gratitude for all of God's blessings
by Betty and Ken Clarke; 22 August 2007

Center: Last Supper
Left: Changing water to Wine
Right: Loaves & Fishes

Photograph courtesy of Betty Romesberg Clarke

Palette and Palates

When Ken Clarke had his brilliant inspiration
To honor those who serve and feed the parish
(Including Betty Clarke), the dedication
Oh-yes!—A stained-glass window seemed just right:
Ecclesiastical, yet cheerful: bright
And dignified, restrained (not brash or garish).
A stained-glass window might seem out of place
With mixers, cutting-boards and stove and sink—
But call it "lab" or "kitchen," the same space
Could well be used to make glass works of art
Or luscious meals that warm the diner's heart.
The two have more in common than you think!
A window lets in light, which, when refracted,
Translates to colors that must then be fixed
In substances like silica, extracted
From … other stuff, which in the right proportions,
Brought to a molten state, subject to torsions,
Is rolled out flat, cut up in bits, then mixed.
Creating fine artistic compositions,
With bits that fit together in designs
To please the eye (or palate)—these ambitions
Apply to both the glazer and also by the cook.
Both work by flair and also by the book
(Though cooks must work in far more rushed conditions).
Thanks, Ken and Betty, on this day of days!
Emmanuel Church will keep the memory green
Of those who serve their church in humble ways.
We may not be the Biggest or the Most,
But we're the only parish that can boast
A stained-glass window in our own Cuisine!

Ruthalma Johnson (author)
Victoria Johnson Brown (editor)

#9 Elaine: Elaine married Wilbur Johnson. They had two children. Cheryl married Roger Harbaugh and they had two children. Ken graduated from the University of Pittsburgh and earned a degree in Law. He, with his wife and two children, live in Somerset where he has a private practice.

My **Sister Elaine**
by E. Jay Romesberg

When I first heard that Elaine had Alzheimer's, I was both angry and sad. It seemed so unfair that she, who had a lifetime of helping others, would now herself be struck down with such a debilitating disease for which there is no cure. We will remember her kindness and generosity and we will miss her. She left us much too soon.

During her life, we remember all the people who were given care by Elaine. She was not paid money for this, but rather she did it because of her kindness. She helped take care of her own parents, her husband's parents and many others. She put other's needs before her own, starting from a young age and continued to do so throughout her life. When she graduated from high school, she had to forgo an opportunity to pursue a career in music, because our Mother was ill and she was needed at home. When she married and had a family, she always put the needs of her husband and children ahead of her own.

When our Mother died, Elaine became in many ways the new Mother in the family. Her house became the central meeting place for out of town visitors. For many years, I came to her house at least once a year. Many times I would arrive late at night and she would have a meal waiting for me. Not just a meal—but it would be something that she knew was a favorite of mine. Many times, Jean and I and Luella would stay at her place even though she did not have the best accommodations. But she had something else. She had the best hospitality. She would insist that Jean and I sleep in her bed. She and Luella would use the couches. Sometimes we had trouble getting to sleep because we would be talking and laughing. We were having fun, as we always did, when we stayed there.

Many of my best lifetime memories are of my times spent with Elaine when she and Luella visited us in California and we would take trips along the coast and to places like Yosemite Valley and Alcatraz. We also took several trips from Pennsylvania to Ohio to visit Floyd. A flood of memories comes to me from these trips. Such as the time we went to lunch at Clint Eastwood's Hogs Breath Inn in Carmel and we were seated next to this new age man with very long hair and who was wearing three or four chains around his neck, had a ring in his nose and one in each ear, also one in his lip and probably some in places we didn't want to know about. Elaine as well as the rest of us had trouble keeping a straight face. Also there was the time we stayed in a hotel suite right on the beach in Oregon and we gave Luella and Elaine the best room which had a big window overlooking the ocean surf. I asked Elaine the next morning how she slept and she said in a humorous way, "Not so good, that darn ocean kept me awake." On these and other trips, especially our several trips to Ohio, we had fun and Elaine was a joy to be with.

I will miss Elaine. My trips to Pennsylvania will never again be the same. But the memories of my time spent with her will continue to make me smile. —E. Jay 8/25/05

#10 Floyd: My career working for the Dow Chemical Company in Midland, Michigan started just three days after receiving my Ph.D. in chemistry from the University of Cincinnati. By then my friend Dick Fulmer, after receiving his Masters Degree in Chemical Engineering at Bucknell, married Jennie's sister, Laura, and worked three years for DuPont in Philadelphia in their plant laboratory. He was then transferred to their plant laboratory located near the G.M. Buick plant in Flint, Michigan, just fifty-five miles south of Midland.

Dick and I decided to go hunting and fishing on weekends. We hunted deer with bow and arrow and pheasants, squirrels and rabbits with gun. We went fishing on lakes and we built an ice shanty to spear perch through the ice in the winter. Perch spearing was our most rewarding event. We froze large quantities of skinned perch.

Dick went back to Delaware for a week each summer to fish with his dad. During this trip in 1965 when he was thirty-eight, he was struck by lightning and died. He was with his daughter, Carol, then twelve. She was unhurt.

My last fishing trip with Dick was forty-three years ago, just two weeks before the accident. On that last trip with Dick, we left some fishing tackle on the west bank of Sanford Lake just north of Midland. We planned on going back after it the weekend after his vacation. I never went back for it. Likely it settled into the sand and to this day marks the spot where two friends shared their lunch together for the last time.

I worked twenty-four years in Midland and then ten years in Granville, Ohio for Dow Chemical Co. I then reluctantly took early retirement and worked an additional fourteen years as a consultant.

At Dow I conducted research and development on fabricated plastic products. Except for patents, none of the work was published. Products I worked on included:

1) Industrial fibers to replace jute
2) Household film products: Saran Warp, Handiwrap® and Ziploc® Bags,
3) Industrial Film: bags, bread wraps, envelope window film, rubber release film, and adhesive film,
4) Metal laminates: auto bodies, telephone cables, and food containers,
5) Composites: Truck and auto head liners and panels
6) Styrofoam®.

We continually worked on product modification, improved quality, process changes and cost reduction. Following are examples of typical projects:

1) Loyd Lefevre, Bob Mathieson and I wrote several patents on industrial fibers made from film for draperies, canvas, jute and burlap replacements, bags and carpet backing.
2) Frank Goff and I wrote a patent on a seal-and-peel film for cereal boxes, etc.
3) Lyle Colburn and I wrote a patent on a film liner for autoclave metal food cans.
4) Lyle Colburn, Ken Bow and I wrote patents on puncture resistant telephone cable.
5) Gerry Schultz (Technical Service) requested a product that used more expensive materials, but would sell at a lower price. Problem was solved by process changes with a significant increase in production capacity.
6) Product became tacky and would not work for customer. Problem solved by changing molecular weight distribution.

7) Product lost adhesive properties. Problem solved by eliminating low molecular weight fraction.

8) Product became too costly. Designed new process and built new plant.

9) Etc, etc.

Beverly Ann was born in 1957 in Midland, Michigan. Pop and Mom came to visit us then. Pop enjoyed going for walks and pushing Beverly in the stroller. Beverly figure-skated and played clarinet and piano. She received the Paderewski Award in piano at age fifteen. She graduated from Michigan with a BS in music. She went to four Rose Bowl games—three with the Marching Band. She had two children: Katherine and Victoria. Both are skilled tennis players. Katy is now a senior at Ashland College, and Torrie is a sophomore in Chemistry at Ashland College.

After raising her girls, she decided to go back to school and work. After studying chemistry at Ashland College, she is now a grad student at Ohio State, in 2009.

Cynthia, born 1964, studied ballet, clarinet, band and figure skating. When in sixth grade, she took first place in clarinet and piano in the 6-8 grade level, sponsored by the Music Guild. She also received the Paderewski Award in piano at age fifteen. She won a Miss Southwest Ohio beauty pageant. She received a Psychology degree at Taylor University where he daughter Ashton was born.

Cynthia first married Christian Robertson. They had two sons, Christian, now studying at Miami University. Their son, Grant, is a junior at Granville High School and plays soccer. Cynthia married a second time. She had two more children: Thaddeus and Ciahnna.

Cynthia now is a sales person at Lowe's and is also a real estate agent.

Floyd Eric, born March 15, 1966.

Updated 9 May 2008

Curriculum Vitae

Floyd Eric Romesberg

Department of Chemistry

Date of Birth: 15 March 1966
Place of Birth: Midland, MI
The Scripps Research Institute, CB262R
10550 N. Torrey Pines Road
LaJolla,CA 92037
Ph: (858) 784-7290
Fx: (858)784-7472
email: floyd@scripps.edu

Research Interests: Using multidisciplinary approaches to understand evolution. Using biological and chemical approaches to understand and combat the evolution of bacterial resistance. Developing an artificial genetic system with which to expand an organism's genetic code. Using femtosecond spectroscopy to understand how proteins are evolved for function.

Laboratory Website: http://www.scripps.edu/chem/romesberg

Education

1994 Ph.D. in Organic Chemistry, Cornell University, Ithaca, NY.

1990 M.S. in Organic Chemistry, Cornell University, Ithaca, NY.

1988 B.S. in Chemistry, Ohio State University, Columbus, OH

Academic and Research Experience

2006-present Associate Professor, Department of Chemistry, The Scripps Research Institute

1998-2006 Assistant Professor, Department of Chemistry, The Scripps Research Institute

1998-1998 NIH Postdoctoral Research Fellow under Peter G. Schultz, Department of Chemistry, UC Berkeley, Berkeley, CA

1994-1994 Ph.D. candidate under David B. Collum, Department of Chemistry, Cornell University, Ithaca, NY

1986-1988 Undergraduate research student under Matt Plate, Department of Chemistry, Ohio State University, Columbus, OH

Honors

2008-present	Member, Institute for Defense Analysis, Defense Science Study Group
2005	World Technology Award Nominee in Biotechnology
2004	Discover Magazine Technology Innovation Award
2004	NSF CAREER Award
2003	Camille Dreyfus Teacher Scholar Award
2003	Susan B. Komen Breast Cancer Foundation Award
2002	The Baxter Foundation Award
1994-1996	NIH National Research Service Award Postdoctoral Fellowship
1987	The Mac Nevin Award

* Eric has 73 publications.

Eric and Jodie live in La Jolla, California with their three-year-old son, Samuel. Jodie is also a Ph.D. with a Post-Doctorate and also works for Scripps Research Institute.

Needless to say, I am a very proud father.

#11 Ephraim: Life After the Farm by E. Jay Romesberg

My high school yearbook claims that my ambition was to leave the farm. I always felt that statement was misleading. It makes it sound like I didn't like the farm. That was not true. I have stated many times that I felt sorry for anyone who didn't have a childhood like mine. On the other hand, when I was a senior, or even before, it was pretty much understood that when I graduated, I would in fact, leave the farm. There were several reasons for this understanding. First, and certainly not the least, I had a bit of a problem with asthma and hay fever. I coughed so much that my brother Merle gave me the nickname "Barkey" because, when I coughed, I sounded like a dog barking. My brothers and Dad gave me a break on farm work by letting me steer clear of dusty jobs. That alone was enough reason to assume that I would leave the farm. Additionally, it was a small farm and there wasn't room for me. So in truth, staying on the farm was not a choice open to me. Nor was staying on some other farm a viable choice. The only question was, where would I go?

I think that I excluded college early in the decision-making process because, at that time, I was not very motivated to go to college, plus it would have been a financial burden for my parents. At that time, I didn't see a clear path to financing the costs myself. Serving some time in the military seemed to offer a clear alternative. There was a very good possibility that the military would provide some good experience as well as some time to mature and decide what to do in the future. Serving in the military would, or might, also provide financial help for college.

So shortly after high school graduation, Al Kusch, Bob Ringer (my nephew) and I decided to join the Navy. Unknown to us at the time, Robert (Red) Engle had made the same decision. We would meet up with him later at the submarine base in New London, Connecticut. All four of us were part of the 1948 Rockwood High School graduating class.

When the time came for us to leave, Bob Ringer and I walked down the road from the old Farmhouse on our way to pick up Al Kusch and then on to the train station which would take us to the Great Lakes Training Facility in Illinois. As we walked down the road, I am not sure that we realized that we were walking away from our childhood. I know now that our mothers did. They both, my mother and my sister Mary (Bob's mother), cried as they watched us leave for the last time as their little boys. I think it was especially difficult for my mother since I was her baby.

Suffice it to say that I never really left the farm. Although from that day that we walked down the road to our new destiny, I have never been back to the old homestead for more than a week or two at a time. I have returned to the area almost every year, sometimes several times in the same year, but only for a day or two per trip. The home place remained my official home address for the five years that I spent in the Navy even though I spent so little time there. After discharge from the Navy, I married Jean Whitehill from Norwich,

Connecticut and established residence elsewhere. Residences for the next fifty-five or so years after discharge from the Navy would include Norwich and Willimantic, Connecticut for four years, while getting an Engineering degree at the University of Connecticut. By the time I graduated, we had the addition to our family of two sons. Following graduation, I accepted a position at General Electric's Knolls Atomic Power Laboratory in Schenectady, New York as a design and test engineer on various systems for the Navy's Nuclear Powered Submarines. During this time period, I obtained a Masters Degree at RPI in Mechanical Engineering with a minor in Nuclear. During that time period our family grew by two giving us a more symmetrical family, i.e. two boys and two girls. After twelve years in Schenectady, I accepted a position with General Electric's Commercial Power Division in San Jose, California and remained with GE until semi-retirement in 1990 and until full retirement in 2003. While working for this division, my job required several moves for relatively short durations. For example, after our initial move of one husband, one wife, two sons, and two daughters to San Jose in 1969, our family just barely got settled in our big house on Mojave Drive when we had to move (husband, wife, four kids, and a dog) to Spain for a year where my assignment was to be the Lead Test Engineer for startup of the Nuclear Power Plant called Nuclenor. We rented our house while we were gone. After Spain we stayed for the summer in San Jose and then moved (same husband, wife, four kids and a dog) to Brattleboro, Vermont, where once again, I was asked to be the Lead Test Engineer for the startup of the Nuclear Power Plant called Vermont Yankee. We rented our house while we were gone. Then in 1975, I was offered the job of GE Site Manager during the construction of two 1100 MWE power plants near Pottstown, Pennsylvania. I accepted the offer and this time we sold the big house on Mojave Drive. We figured that we would not come back to San Jose. Our two daughters made the move with us and the two sons planned to follow later. Bad decision. Due to a health issue, I had to come back to San Jose. As a result we had to buy another house since we had sold the big house on Mojave Drive.

During our one-year absence, house prices had nearly doubled in San Jose. So financially, we had made a bad deal.

After that we stayed in San Jose for a little while until in 1979, I was asked to take the Lead Engineers role at Corso in Italy because the guy that was there decided he wanted out. He claimed to be worried about threats that were coming from the Red Brigade. I went by myself this time. I made a number of trips back and forth during my total stay of about one year.

For the remainder of my GE days, I remained in San Jose as my main location and made numerous trips which did not require moving the family.

During all of these relocations I have never gotten very far from the farm. Many times I have thought that I might have been better off if I had stayed closer to home, but then I remembered something that my daughter Tricia said to me one time. They were words of wisdom and I never forgot them. She said, "You should never say, 'I should have'." I realized that she was right. It is a very useless statement because you are wishing that something you did in the past could be changed. It is impossible. But after thinking about it, I decided that there is, in my opinion, an absolute truth in life. It is this: We all encounter many forks in the road of life and when we do, we have to decide which fork to take. As Yogi Berra would say, "When you come to a fork in the road, take it." Well, that's easy for Yogi to say. But many forks require a lot of thought, others are very simple. But nevertheless, we have to decide. Well, here is my absolute truth. At least it is for me. Up to this moment in time, every time that I came to a fork in the road in the past and I had to decide which fork to take, I HAVE MADE THE CORRECT DECISION. Some of you may disagree, but think about it. Suppose, I had decided not to accept the job offer in San Jose, and we had stayed in New York. Who would my kids have married and what would my grandkids look like. If we had stayed, maybe a tree would have fallen on my head. Maybe life would have been better. Or maybe it would have been much worse. Would I want to find out if by some magic I could go back and choose the other fork? Certainly not. Would you want to choose a different fork? Perhaps if you are sitting on death row, you might take a different fork if you could. But I think most of us would agree that every fork we took up to this point in time was in fact the correct one.

Footnote:

> At the time of this writing (summer of 2008) Jean and I plus all of our direct descendants which includes two sons, two daughters, five grandsons, two granddaughters, and one great-grandson all remain in California. When we have family gatherings which include significant others, we usually have a total of about seventeen people. So out of two, comes many (I heard that someplace before).

I haven't checked this out thoroughly, but I think that I might be the first Pennsylvania Romesberg who has made California home. So I guess we have started a dynasty in California.

Gary, our oldest at age fifty-three, lives about one marathon (twenty-six miles) south of us in Morgan Hill and currently is employed in San Jose where he is in charge of Medical Imaging (i.e. MRIs, CAT scans, x-rays, etc). He has become a gun collector and with his dog, Bo, he loves to hunt ducks and other water fowl. He and Marie were divorced some years ago and she lives in Gilroy just south of Morgan Hill. They have three sons. The oldest is Austin who is married and living in Fresno (about 100 miles south of us) and is working full-time in a managerial position while going to school to get his degree in Business Administration. The next son is Callan, who is also married and has a son (my great grandson) and they also live in Morgan Hill. Callan put further education beyond high school on the back burner to provide for his family. His business card says he is a Service Writer for Precision Tune Auto Care. Gary's third son, Jason, is still in high school and says that he wants to join the Marines. All of Gary's boys were on the high school wrestling team and played some basketball and football.

Tom, our second oldest, lives in Bakersfield (home of Buck Owens), California with his wife, Carla, and son, Ephraim, (named after me, I guess). Tom followed in my footsteps in college and got a degree in Mechanical Engineering. I tell people that he followed in my footsteps, but when the footsteps ended, he kept on going. He has been General Manager of an 1100-megawatt gas-fired power plant called LaPaloma for several years, plus he has other corporate responsibility relating to power generation. He has a huge workshop in his back yard where he can make or fix just about anything. He also has a huge garden. He has a Harley Davis motorcycle, which he sometimes rides to work. Carla is a part-time nurse and Ephraim is in fifth grade and is very interested in science, including the periodic table, and bugs. Ephraim is somewhat of a loner, but his grades have been essentially straight A's.

Patricia (we call her Tricia) is married to Bryan Wing and they have two daughters and one son. They live in San Jose about twenty minutes from our house. Bryan is a manager for a Silicon Valley Company and has been a very good Daddy to his three kids. Their oldest daughter, Emily, is an avid reader (she has read more books in her short life time than I have read in my whole life). She is graduating from High School this year and is currently planning to attend college. Olivia is in Junior High School and is becoming a very good piano player and dancer, as well as an excellent student. Graham is in 4th grade and I often call on him to solve Computer, DVD, or TV issues because of his working knowledge of all the modern electronic gadgets.

Laura lives about 1.5 miles from our house and teaches science in a middle school near by. She coaches the cross country team in addition to her regular classroom responsibilities. Sometimes she gets me to run (or walk) with some of the team members. Laura has been my best supporter and attendant during many of my ultra distant marathons. She was with me when I ran the Mother Road 100-miler in Oklahoma when I was just two weeks shy of my 76th birthday. I had problems late in the race and had it not been for her, I would have dropped out at mile 95 where I was having major problems of just standing upright.

Some years ago I met an ultra-distance runner named Joanne Sliger. She told me that she was the fifth generation in her family line born in California. I figured that her first ancestors must have come here with the Donner Party or at least in or near that time frame. All of my grandchildren were born in California which makes them first generation of my descendants born in California. Callan's son C.J. becomes the second generation born in California. Amazing, how fast that happened. —EJR 2008

Chapter 16

Memoirs

The Ephraim Romesberg Family—1947
(Photograph from family records.)

Memories of the Early Years
by Luella Romesberg Ogle

I was born Monday, August 31, 1914. Mom said she ate a big noon meal of corn on the cob and then I came. I can remember when the doctor came to the house to vaccinate me to go to school and how frightened I was.

Happiness was when we moved in the spring time from the big house to the summer house. Mom cleaned and polished the kitchen coal and wood stove in the big house with kerosene and put newspapers on top of it. She cooked thru the hot summer on the coal and wood stove in the summer house.

Mom cooked three big meals every day. At the noon meal we always had hired men to feed—plus Uncle Alex. Several summers, black men worked the fields. There was Sammy, Soko and a third whose name I can't remember. They were friendly men who came from France.

Nobody ever stopped at our house at meal time that Pop didn't invite to stay and eat. Mom was always pregnant or toting a baby while cooking.

Once I stepped on a baby chick—I can still feel the squish under my foot.

And Wilbur was a pistol. He threw the inside works of a clock up into the air and it came down and cut him on the head. He told Mom I hit him. The doctor came to the house one day to clean his ears. Wilbur had put sand in them.

When Mom went to the barn to milk the cows, I had to sit by the cradle and rock the baby. We always had a baby to care for. When I washed dishes and couldn't reach the table, I kneeled on a mooly chair (a chair without a back).

Uncle Alex always sat in the kitchen humming with his chair tilted back. I don't remember him ever working. He had a mustache and, as he slurped his coffee, he made a strange noise. I think that he never changed his socks. He was a quiet man. He was engaged to be married but the girl died two weeks before the wedding. I always thought Uncle Alex mourned that all his life.

Our house had six rooms when I was little. The kitchen, sitting room, and Gramma's bedroom were downstairs. Uncle Alex's bedroom was at the top of the stairs. Mom, Pop and all the kids, I think five of us at that time, all slept in the middle bedroom in two large beds. In the end bedroom, two teachers who boarded with us for three yeas, Elizabeth Hoose and Mary Swearman and the hired girl slept in one bed. For a while Gladys Weyant worked for us and then Bessie Kelly.

The house was pretty when I was little. There was striped red and green carpet through the upstairs and on the stair steps. The kitchen was papered with pretty oil cloth and the woodwork was painted black.

Gramma Romesberg spent most of the time with us. Mom was always having babies and needed her. Gramma was a super Christian lady. I remember her going behind the smoke house to pee. She spanked me once. When she came to our house one time for a visit, she brought yard goods for Aunt Della to make a dress for me. She said, "Be sure and have it made with long sleeves." Gramma died when I was a sophomore in high school. She died at Uncle Joel's house. I went to see her. I stood in the doorway sobbing and crying while Reverend Miller was holding Gramma's hand and praying. She lifted her other hand and waved to me. I have pleasant memories of Gramma.

One time one of the kids turned up missing. I think it was Betty. We searched the neighborhood and I went up to Uncle Milt's and looked in the spring. I was so scared. Then we found her behind the cradle, fast asleep on the floor. One time one of the kids salt and peppered the eyes of the baby sleeping in the cradle. That old cradle holds many memories. Gramma was rocked in it. Wayne has the cradle now.

In the summertime, Merle and I peddled milk in gallon buckets to Wilsoncreek in our bare feet. Mom had a few regular milk customers. We were scared to walk through the woods and ran most of the way.

Mom would churn butter in the spring house and sing "Abide with Me" at the top of her voice. She always had sore ulcerated legs.

Some Sundays in the summertime, Uncle Joe would take us in his two-seated Ford to visit our aunts and uncles in Summit Mills— Mom's family. Sometimes we went on Saturday in the horse and buggy and stayed over night to visit Gramma and Grandpa Swearman. Merle and I used to ride in the back in the luggage carrier.

Pop and Alex talked Pennsylvania Dutch most of the time. When Pop didn't want us kids to understand, he would talk in Dutch to Mom. Pop only went as far as the third grade, but he taught himself to read and write. He played the coronet in the Wilsoncreek Band and in the evenings he would sing and play hymns.

Mom kept having kids and the house became too small. Pop and Uncle Joe dug a big hole on the south side of the house, then built a basement and moved the old six-room house onto the new foundation. Then he and Uncle Joe built four large rooms on top of the old foundation, two up and two down. I can still smell the new plaster. Things stood still when the depression came and the rooms didn't get wallpapered and painted for years.

We got electricity when I was a junior in high school. How exciting when the lights were turned on. Before electric we had kerosene lamps until Pop installed a carbide system. There was a huge tank in the yard that held 100 pounds of carbide. It was piped into the lights that were pretty wall fixtures. One time when Pop and the boys were dropping the five-pound emergency carbide into the tank there was an explosion. Merle's face was burned—he cried most of the night. Pop pow-wowed for the burns. People used to come to our house from miles around for Pop to pow-wow for problems. I remember Pop moved his hands like a wand over the problem and said words to himself. He wasn't allowed to reveal the words or the power wouldn't work.

We had a hand pump in the kitchen and pumped fresh, clear, cool water from the spring house. There was a bucket in the sink always full of water and we all drank from the same dipper.

The mattresses on the kid's beds were big ticks, made the size of a regular mattress and filled with straw. Every spring and fall Mom would empty them and fill them with fresh straw.

I remember house cleaning time. The beds were all torn apart, the room was emptied and everything scrubbed with Mom's home-made lye soap. We had to dunk the bed slats in a bucket of water, first one end and then the other. The bed springs were also washed.

Every Saturday morning, one of my jobs was to take the cooking kettles to the ash pile and scour the bottoms with ashes. Mom would always scold me if I didn't clean the ears of the cooking pots.

Our dish cloth was a sugar sack or some sort of a rag which was always washed separately—never with the other clothes.

Babe, a horse, fell and broke his leg. I can still hear Pop cry when he had to shoot the horse.

Sister Della died October 10, 1925, at the age of six from typhoid fever and pneumonia. Mom burned sulfur in a dish in every room to fumigate the house. She burned the big sticks of O-Boy gum friends had bought for Della. I was heart-broken when she burned the gum. Della had eczema all over her body, mostly on her hands and arms. She scratched until it bled. Mom made long white muslin mitts and tied them on her arms. The night before she died the pot belly stove in the sitting room wailed and Pop cried when the Confluence Hospital called and said that they feared that she wouldn't live. They brought the little casket in the front door and put it in Gramma's little bedroom. It was a very gloomy, snowy day. Teddy, our dog, walked around so sad with his head between his legs. People brought food to our house and I remember a big can of store-bought peaches. I was eleven years old.

Gramma Swearman died when I was thirteen. Cousin Howard took Mom and the two youngest kids over to Summit Mills to Gramma's house. On that occasion, I took over the house and I wanted to bake pies. I mixed up a big batch of flour and lard but I couldn't roll it out for a pie. It crumbled all to pieces. I took the bowl of pastry and hid it in Mom's fruit cupboard in the basement. Later, when we came home after the funeral, I told Mom about it. She got the bowl of pastry and was so kind and nice to me. She said, "It's all right; all we have to do is add more flour and cold water." I think they were the best pies Mom ever made. Today, pie baking is one of my favorite cooking jobs.

I used to spend a week or two every summer at Gramma Swearman's in Summit Mills. She always baked a lot of pies. Her basement had white-washed walls (lime mixed with water) and a ground floor. It was nice and clean. I can still smell that basement. I can still see the house and remember the furniture, the big front porch with Grand Pappy sitting there smoking his pipe and Gramma's woven rugs all over the floors. Gramma always had a baseball for me made out of woven strings. She had a big goiter protruding from her neck. She was only fifty-seven when she died from a cerebral hemorrhage.

One time Merle, Wilbur and I had spoons and we were "painting" the house with mud back in the corner by the chimney, just as far as we could reach. Pop came around the corner, doubled over, bit his tongue and started after us. Merle and Wilbur got away. He caught me and paddled me. To this day I owe those two boys because they escaped the paddling.

We all had to work. Pop always said I was his best potato planter. He praised me and then I tried to do better—smart psychology.

When Mom got pregnant with E. Jay, she fell down the stair steps and broke her collar bone. I had to take care of the house. I was fifteen.

E. Jay was born early one morning and afterwards I got a ride to school with the doctor. It was the last day before a two-week school vacation and then I was home to do the cooking and take care of the family. Thank goodness he was the last baby.

I was with my friend, Geneva, one night, walking along the path to Wilsoncreek when we found part of a pack of cigarettes. We sat on the fence in the cow pasture trying to smoke them. I came home and later, sitting in the kitchen, Mom said, "I smell something burning." When she said it the second time, I high-tailed it off to bed.

We used to sit around the kitchen table at night, eat apples and draw and color pictures. Mom was lots of fun. Pop always went to bed early.

A big, fat lady with lots of bright colored skirts and dressed like a gypsy came to our house and washed clothes for us with a wash board in a tub. Her husband came along and sat by watching and smoking his pipe.

Mom also had a neat, skinny, black lady who came to do house work. One time she took me home with her to Blackfield. She had a very clean, bare house. Her bare floors were scrubbed almost white. She gave me a baking powder biscuit.

Eliza, a jolly, fat neighbor came often to help Mom. She took produce or dairy products for her pay.

Pop always took us to Sunday school. We walked the railroad tracks to Sanner Lutheran Church. I can still feel Pop's big hand holding mine as we crossed the railroad trestle.

Uncle Milt took Merle, Wilbur and I to a circus in Rockwood one time. Pop gave him money for us to spend.

Merle worked hard on the farm. We all had our special place at the dinner table. Alex at the one end, Pop at the head. I used to fight with Merle every meal because he smelled like the horses and cows. Now I am ashamed of picking on him. Nobody was allowed to start eating until everyone was at the table. Kids lined up behind the table on a long bench. Pop always asked the blessing and we used to tell him he prayed too long because the food would get cold. We were not allowed to talk with our mouth full or to drink anything until our mouth was empty. We always had good food. Nobody can make vegetable soup like Mom used to make.

You haven't lived until you've experienced running through a cow pasture in your bare feet and tramping onto a pile of warm cow manure with it oozing up between your toes!

When I was little, all of us kids would line up to watch when Pop killed a calf for butchering. He would hit the calf on the head with a hammer and then he would slash its throat.

I always had a play house in the corn crib during the summer.

I cut one of the baby's hair one time straight across the back. Pop took one look and said, "Why, Luella, if you can cut hair that good, you are going to cut hair for all of us." He sent to Sears Roebuck for a barber outfit and I became the family barber. Elaine will vouch for me being a great hair cutter. I gave her a boyish bob one time, cut off all her pretty curls. She didn't want to go to school and hid behind a tree.

Mom wouldn't allow me to use the sewing machine. One day when she was out helping to haul hay, I sneaked a piece of fabric and made Helen a dress. When she saw what I could do, well, from then on I was allowed to use the sewing machine. She also would never allow me to milk a cow, said I would spoil the cow. Then Betty came along and became a milker.

Pop smoked and cured our meat in the smoke house and how good it smelled! The hams were smoked and then buried in the grainery in the oats bin. Mom canned all the other meat.

On butchering day, meat was cooked in a big iron kettle to get ground for the pudding meat. Mom cleaned and scraped the hog intestines to make the sausage.

We loved butchering day and, when we lived in Boswell, we always went home on that day. John would take off work for that and also for threshing day. We never missed these events, especially the big dinner Mom prepared.

I walked to Rockwood High School for four years, usually hitched a ride. Mom said she used to stand at the window and cry when I walked through the field on snowy, cold days.

One time she called the school and asked me to bring some fuses home. Floyd had caught his arm in the washing machine wringer and blew a fuse.

We paid $2.00 a month for the telephone. Our ring was two longs and three shorts. When they raised the bill to $2.50, Pop had the phone taken out.

I graduated from high school in 1932 during the depression. Pop gave me $25.00. Mom went with me to Somerset and I bought a coat, dress, shoes, purse, hose, underwear and a hat all for $25.00 at J.C. Pennys.

Woodrow and I married right after graduation. Pop was angry and the day we ran away to Cumberland, Maryland to get married, he had a terrible migraine. We went to live at Pappy Matts', no job and no money. Peggy was born there. Pop came to see us and said, "I want to name this baby after my Aunt Peggy. I forgot to name one of my girls after her." So we named her Peggy Jean. We lived across the field from Pappy Matts, where Chick and Minnie Romesberg later lived. JoAnn was born there three-and-a-half years later.

Then tragically, Woodrow died in April and left me with two little girls. Still no money, but two sets of parents who took care of all the bills. The doctor said Woodrow had acute hepatitis. Now his sisters tell me that he had leukemia. The girls and I stayed at Pappy Matt's and Gramma Lil's house, Woodrow's parents.

John Ogle came one day when I was tending the gas station. Not being used to the new 1937 cars, I spilled gasoline all over his white shoes. That's how I met John. He spent a lot of time at his Aunt Mag's, across the field. One night he took Boots, Woodrow's sister, out to a square dance and she coaxed me into going along. That's how I got hitched to John.

First written by Luella in March 1989.

Some Things I Remember
by Paul Romesberg

Growing up as a family on the farm.

I remember Merle plowing with the horses on the Long farm. I was small and we got to laughing about something and we both fell on the ground and the horses kept going and the plow stayed up. We couldn't say "Whoa" to the horses.

I also remember Merle and I going up to Luella's for her husband Woodrow to cut our hair and going home at night—it was so very dark. We heard a noise and ran fast, afraid something was coming after us. It was just a rabbit.

I remember going hunting with Merle, Wilbur and Dad when I was sixteen. I wanted to go with the boys and they made me go with Dad. We went out above Matt Romesberg's place. Merle and Wilbur went one way and Dad and I went back the Dinky track. Dad said he would stand there at an opening and I was to go meet the boys. I only went a short distance and I saw a buck lying alongside a stump and I shot it with my single barrel shotgun and a slug.

Floyd and I went hunting deer at the Ridge in the snow. It was easy tracking and I tried to get Floyd to go home. We ended up on the Schaffer farm in the dark.

I remember Floyd standing on a chair and Mother was using the wringer. Somehow he got his hand caught in the wringer up to his elbow and it pulled him down into the washtub of water.

I remember Floyd starting a fire on the wall behind the cook stove. Then, too, I remember Junior falling off a load of hay and breaking his arm.

Harold and Bill, our first cousins on Mom's side, came over to hunt squirrels at our place and we in turn went over to Silver Valley to hunt rabbits with them.

I remember Mother going to the old barn with a lantern to milk cows. I also remember Grandma Romesberg in the Summer House, which was along the road in front of the house. Later we moved it back into the yard.

I remember the time we had the horses hitched to the bob-sled and it was snowy and icy. One horse fell across the tongue and what a time we had to unhook them without any mishap to the horses or us!

We had many a great time growing up together. We had the best Mother and Dad who cared and took care of us. I am so glad I was, and still am, a part of the family. Let's keep in touch with each other.

I appreciate each one of the family, including nieces, nephews, uncles, aunts, and the whole Clan of Romesbergs.

"Your Brother, Paul"

Some Memories
from Betty Romesberg Clarke

A Player Piano—Fun

An old phonograph player in the attic!

Mom and Pop shopping in Somerset on Christmas Eve with $10 to buy for the whole family. I got a little doll (unclothed) and was so thrilled on Christmas morning that I actually got sick.

I was named by Grandma Romesberg.

Cleaning chicken for Sunday dinner. I chopped off the head (ugh). Mom made noodles every Saturday and left the larger circular plate drying on the dining room table! The dinners were delicious and we always had home-cured ham. It was stored in a bin of oats because mice could not go into the grain to get at it. It was delicious! We always had Sunday afternoon visitors (Mom's sister, Aunt Ruth). They stayed for an afternoon snack or more and always went home with milk, cream or something special!

I remember making cider and our sneaking a taste from the barrel as it aged and it wasn't too hard!

We made sauerkraut and periodically pushed down the plate on top to squeeze; essential to marinating.

Making apple butter, pear butter, and maple syrup. Sometimes a bit of taffy or brown sugar. Delicious!

Weekend beans baking on the back of the stove—also making cottage cheese which we always ate with maple syrup.

Always a box of crackers on the top of the stove—called a warming shelf.

Uncle Alex came at least once a week and sat all day in the rocking chair in the kitchen—never moving when we swept or scrubbed around him. He and Pappy spoke only German and he ignored us. (I think he missed never having a family of his own.) He had a small house in Wilsoncreek. He hummed all the time!

Uncle Milt's hill joined our property. After rain we would look for mushrooms. Once I found one and Mom cooked it for me. It had to have a ring around its stem—that was safe. Once in the winter there was a crust on the snow and I sledded all the way down to the road. It was exhilaratingly special and I alone in the cold night—starry sky. I used to ask Mom if I could go out sledding. Once Elaine and I sledded to Wilsoncreek (a little sled) and saw a big light in the woods which happened to be a house burning down.

There was always a mystery about a grave on the line between our property and Uncle Milt's. Someone was buried there and we thought we saw a ghost which was, in fact, a pathway for people crossing from Blackfield and holding a lantern.

Uncle Milt had a general store in Wilsoncreek and we traded eggs for groceries. He also delivered salt and flour in big bags.

Butchering day was a big event. Uncle Joel came to help and always sneaked off to have a snort of something he hid in the barn. There was a huge iron kettle to cook things to make a pudding or crackles (we called them) and we washed the intestines to stuff and make sausages. These were hung in the smoke house and smelled good all year or as long as they lasted!

We had a routine and it was obeyed. Every Monday wash (Sunday we got water from the spring and put it in a large pan to heat on the back of the stove (no electricity). Tuesday we ironed after drying the clothes outside on a long line—lots of clothes. Wednesday (I don't remember—maybe Thursday). Friday we cleaned upstairs and Saturday downstairs—plus sweeping the outside walks. Saturday was baking day.

In summer we moved our kitchen to the summer house (fun). This left the big house cool. We were cramped in the little summer home, but, all meals were served there.

I remember one salesman who came once a year and sold a cream (Watkins) which cured everything!

Mom lived by the Farmers Almanac!

At Christmas we received a treat—a little bag of hard candy and an orange and a mint too, I think.

No doors were ever locked! I can still hear the frogs croaking. Somewhere the crickets and the dogs barking—we were never afraid.

I often wandered alone in the woods—quiet and not afraid. Remember finding huge moths—filmy and never seen again. Also, once an orchid and sometimes, honeysuckle.

In the winter we had to take the cows to the watering trough to drink—no pasture.

Also Pappy took some of us to the woods to pick out a Christmas tree! I stepped in the creek and came home with a wet foot one time.

We had a path to go through the woods to Milt's store!

We had a sand pile and went there to scour the pans. If we missed around the handle, Mom sent us back.

Once E. Jay was playing in the coal shed (the furnace took coal) and the door jammed. I panicked until I found him. He was black from playing in the coal. I loved him and looked after him.

It was hard work—living on the farm. We planted cabbage and tomato plants and then had to carry water (not very close to the house) and tend the plants.

I think the boys were ornery. Once they brought home a snake and had it crawling around on the floor. They also brought home a possum! I think I had and still do have nightmares from that experience.

We walked everywhere and daily: To Milt's store through the woods and to Luella's and to the fields on the farm.

Family reunions were held every summer. The Romesberg and the Swearman (Mom's). Pap was a frustrated entertainer and to Mom's embarrassment always got on the stage to try to entertain.

Ice cream was like ambrosia. When we found ice in the watering trough, we made it and what an occasion!

A popular game we played at school (during recess) was andy-andy over. We would throw a ball over the school house or the coal house to someone on the other side (unseen) and yell "andy andy over." If you caught the ball, you got a point. We also had to take turns to walk to the nearest house to get a pail of water which we shared and all drank from the same dipper (ugh). There was a coal stove in the middle of the room to heat the place. Wet coats were hung around it and I can still smell them drying!

I think we went home for lunch and sometimes Uncle Milt picked us up on his way home to lunch. I remember standing on the running board.

Well, the outhouse was something else. A long cold walk—it was next to the muddy pigpen and had a Sears catalog for paper.

Sometimes Mom sent us to Aunt Emma's (Uncle Milt's house) to borrow a cup of something—probably sugar. She always had it. Their house was beautiful to us.

One very poignant memory is riding the big old work horse for Pappy to plow. When finished, I had to take the horse home (from the Long Farm) and I was scared. When we came down to the watering trough, he automatically took a drink and I managed to stay on—pretty scared.

We had a wonderful vegetable garden—always a bit of rhubarb and raspberries and then the usual vegetables. We buried apples in the ground in winter.

When potatoes were ready, I remember going out to dig them up. I can still see the plant and surrounding potatoes very well. Seems I was the only girl who got picked to work outside with the men.

We also hunted dandelions in season and Mom would make a hot bacon dressing. Very good!

I rode to Rockwood with Pop in the horse-drawn buggy to get my vaccination for school. I was five and allowed to go because my birthday was in December. I can still see some of that trip. What a job for Pappy—we never appreciated it!

Aunt Della's coming occasionally on Sunday afternoon with her collection of Comics! What a treat. We read them over and over in the yard. She was a lovely lady.

We ate outside often on a big table—especially for the butchers and the threshers—always a big delicious meal. Then we were in the summer house—it was later moved and then accidentally burned down.

I learned how to shoot a .22 rifle. Floyd remembers but I barely remember.

I remember mowing the big lawn with a little mower and it took forever.

I loved E. Jay and rocked him in the kitchen rocker. He had asthma and often had trouble sleeping. Then he would call me.

The pantry had a pitcher pump to get water up from the spring house. I remember pumping and pumping. It was an old iron thing!

Many strangers passed in the evening and Pappy also let them sleep on the porch. We were a bit scared but got used to it. They were "honkies" (men from Europe who worked in coal mines) if you read the Johnstown flood—passing to Blackfield.

There was a large potato bin in the cellar and the potatoes lasted until the new yield in September or October. It was a dark, cool corner so they kept well. We ate potatoes every day—probably several times. Mom baked bread twice a week and sometimes little buns which we loved. We had an early supper and no one snacked after. Some Saturdays someone would bring a gallon of ice cream from Somerset and we had a most enjoyable feast.

Betty

Memoirs

by Lois Elaine Romesberg Johnson

I was born March 11, 1925, the ninth sibling of a family of eleven.

I can't remember too much of my younger years. The earliest picture I have of myself is when I was in third grade. Some of those days are more vivid. I see myself now sitting at my desk with carving next to the ink well that one of the boys carved with a pocket knife. I had a coat on that Aunt Della had "fixed over" for me, Mom said. Those days at Wilsoncreek School were quite different than those that our grandchildren are having today. We walked each day through rain, wind, storms and snow. The snow was very deep at times. Since there were usually four or five of us in school at one time for several years, the oldest boy would wade ahead in the snow and make a path for the rest of us to follow. There had been a few times Pop would hitch a big log to the horse and drag a path for us to walk. We would get cold as we didn't have boots, gloves, coats, etc, as do today's school children. We missed very few days of school and were never late.

We at first had the two rooms, but it was not too long until only one was needed. The seats were straight-backed and at times we sat two to a seat, especially if the school was cold or if the teacher was teaching or reading something really interesting or preparing us for an exam.

The discipline was good, the teaching was well prepared. The older students would take turns in cleaning the chalk board, sweeping the floor, carrying buckets of coal, and also buckets of water from the nearest neighbor. The coal was used to make fire in a pot-bellied stove, which was our source of heat, and the water was poured into a fountain and that was our drinking water. We all used the same tin cup. We had many memories of those days in our little country setting in Wilsoncreek. We had a Christmas program each year. Pop would come at times to see and hear us. One year I wore someone's big shoes.

Betty and I would run down the road, never wanting to be late. Betty usually had her knees marked from falling so often. We often took our lunches and when it was nice, we ate outside on a big rock. I see myself now, there eating a piece of huckleberry pie. I always liked playing prisoner's bases and sometimes softball until one time one of the boys ran into me and knocked me out. I had my "Shirley Temple" dress on, too. I was more worried about my dress getting torn or dirty than I was of being hurt. One by one, we each took our eighth grade exam and passed it in order to go on to high school in Rockwood.

Through our elementary years and high school we had our jobs on the farm. We always had cows to milk. It seemed I couldn't milk too well. Betty did a lot of milking. We raised our beef and pork and chickens too, so we had the eggs along with meat for use. I remember the day when Floyd and I were to go to the chicken house and gather the eggs. We tried carrying them in our hands and Floyd dropped some of his, but Mom wasn't too unhappy with both of us.

We separated the milk in the separator in order to get cream to make butter, cottage cheese and "goodies" like ice cream. Mom usually churned until we kids got older. She would sing and churn. To determine the time to cook a "soft boiled" egg, she would sing one verse of "Abide with Me."

We grew our vegetables in a garden and large truck-patch. We canned a few hundred jars of many different vegetables and fruits each summer. They were from pint to half-gallon jars. We grew berries too and various apples and pear trees. I remember so well the "pippen tree" at the end of the walk. Floyd and E. Jay were in this tree one day and one of them caught their boot on a limb and tore it and it hung there. One of them also threw an apple and broke a sun porch window.

It seems we all had our "nixy" times. Floyd started two fires in the house. Helen was always quiet. I cut her once with a broken bottle on her leg. I was so upset one Easter because I couldn't help dye the eggs. The oldest one was to take over and do the supervising and this was her year. I confided in Helen quite often. On my birthday one year when I belonged to the "Uncle Tammy" club, she had a hanky for me under my plate at supper time. One day I hid one of Wilbur's boots upstairs behind the dresser. He couldn't go down to Uncle Milt's store. Paul would be found at times in the barn or pig pen sleeping. Could have been the chores he was to do weren't appealing.

I never knew sister Della. I was born on March 11 and she died October 10. Mary, who Mom had by first husband, died April 25, 1986. Mary moved some twenty-five to twenty-eight times during her married years. It never seemed to bother her, and she could always try to lighten anyone else's problems. E. Jay wandered away quite often when a little boy. One evening we were looking for him and later I found him over at Harrison Pritts', sitting at the table with a big tin cup of coffee. Floyd would go off too and follow Merle when he was plowing in the field with the horse and walking plow.

Quite often we had, what we called then, a "bum" stop in at our house. Pop would always invite them to come in to have a cup of water from the spring. I can see a few of them now eating a piece of homemade bread with apple butter on it.

We had many farmers in the area. The largest percentage were of the middle class. They grew by clearing land, cutting trees and brush. There was much bartering, paying doctor and dentist bills by giving eggs, butter, milk, sometimes chickens or ham, or a load of wood or chips. Occasionally one farmer would tear down another one's fence or open his railing to let the horses or cows out onto the other pasture or grain field as a way of revenge or jealousy. There were a few shootings and murders. The first man hung in Somerset County is buried on our home place. Trading seemed to be a natural thing. I would walk occasionally to a neighbor to borrow an item, but often Mom and she would just trade whatever was needed. One day I walked up to Luella's to take her milk. This was when she lived near Poppy Matt. On the way home I saw this pretty black and white animal in the road. I kept pushing it with my shoe down to the house—later to discover this was a skunk. My shoe smelled like skunk!

As early as 1905, the citizens of Wilsoncreek petitioned the township board of directors three times to build a suitable building to conduct classes. They were always turned down; finally the citizens took matters in their own hands. One year after that in ten days time they got sufficient funds to maintain an independent school. It was a success and some forty students attended. They built a new one in 1907 after the order of the Somerset court went into effect. Wilsoncreek had a good population then. Uncle Milt's store was the only one I knew. There had been another, as well as a hotel, several churches, dance hall, barber shop, fire hydrant and wooden sidewalks. The mines had done really well. Merle and Wilbur shoveled-up coal from along the tracks or banks. They later used the horse and wagon to haul loads to be used for fuel. Many roads and by-roads became reality. It took lots of chopping, pushing, digging and sweating to gives us a clear path or road to walk to school, Sunday school and to visit our neighbors. So much had to be done by hand until the progress of machinery. Pop spent long hours cutting timber and logs, and later he was the foreman for the electric company, cutting out brush for electric lines. Merle and Wilbur worked for him and walked several miles daily. They cut posts and sold them to help buy feed, grocery and clothing. Floyd helped cut posts and sold them to pay for his college tuition.

Blackfield was a thriving little town with a school, grocery store, meat market, bowling alley, church, boarding house, pool room, and the mining business was good. In November of 1916, there were sixty-eight pupils in the school with one teacher, Luella Brant. They found it necessary to have some of the students come mornings and the rest afternoons.

I went to Sunday school a few times bare-footed, but that wasn't unusual. Sundays were the days for company—usually the aunts or uncles. Aunt Della came often and she usually had a dress she sewed for Betty and me. She and Uncle Scott would take us along back to Acosta for our summer vacation of a week. Mom always had home-made pies or cookies for the Sunday company. I remember one day she had a few pies cooling on the wooden sink by the window, and Wilbur, a neighbor boy and maybe Merle or Paul sneaked one of them and took it down to the woods and ate it.

The years passed by and there were times of much joy and also pain. Because of being so much younger, Luella always seemed so grown-up to me. She always wrote such good poetry. The day she cut my hair above my ears and straight across I was mad at her, but I forgave her. Merle and Helen each lost a finger accidentally and I was so scared. Paul and Merle spent times in the service during World War II. Wilbur had many ordeals, the death of Doris, his first wife, and son George, much surgery, and many hospital stays. The loss of Helen's husband, John, caused her to have to make many decisions. Betty has now been traveling extensively since living in Switzerland. Floyd caught his arm in the washing machine wringer. E. Jay had many bouts of asthma, but through all the sacrifices, pain and heartache, came the reality of two prosperous farmers, a nurse, dental hygienist, a chemical and mechanical engineer, as well as a preacher in our family.

I married Wilbur Johnson on October 10th, 1946 and he died of heart failure January 8th, 1981.

March 1989

JoAnn's Memories of Eph's Farm

Getting ready to go to Grandma and Grandpa's back in the 1940s was always exciting. Dad had a '36 Chevy and he and Mom and my sister Peggy and baby brother Johnny would pile into the car in Erie and head for the farm. Back then it was an all-day trip. Mom would pack up food and we would eat along the way as there were no fast foods back then. We would hardly be out of Erie before one of us had to go to the bathroom or get hungry and I usually got car sick. Dad would sing songs, like "The Bear Went Over the Mountain" or "Clementine." We also played count the cows on your side of the road and a church added ten points and a cemetery took away ten. Whoever got to a hundred first won. Dad would slow down for his side and speed up for the other. Then we had the Burma Shave signs to read and sometimes we stopped in Pittsburg for a good ice cream cone. We would arrive tired and dusty as many of the roads were un-paved, but we were always happy to get there and they were always happy to see us.

My Grandfather Eph was a man I always looked up to and I feel that everyone who ever knew him respected him. He was a God-fearing man and could quote verses from the Bible at will. I believe he was a fair man. One of my earliest memories of him is when I rode on the buckboard beside him to go to Uncle Milt's store for a bag of feed or flour. Grandma had asked him to pick out a blue flower print. I must have been six or seven years old, and Mother dressed me up in Sunday clothes and a hat and it was just Grandpa and me. I felt very important to ride beside him on that horse-drawn buckboard. When we were in the store, Grandpa called me over to where he and Uncle Milt were and there were two bags of blue flower print and Grandpa asked me which one I liked the best and I remember the cornflower blue print I picked and he gave me a penny for choosing one. I held that penny between my right thumb and forefinger. It was precious to me, and with it I picked out three pieces of candy! Not long ago, I found a penny and picked it up and happened to hold it again between my thumb and forefinger and it brought back that memory of a time when a penny could buy three pieces of candy from the candy counter.

My Grandmother Mayme was always fun. When I look back on how much work she did in a day, I wonder how she kept such a good sense of humor, but then I guess that helped her get through the day. She was always working in her huge kitchen, braiding rags into rugs, or shelling peas or snapping beans or churning butter or baking bread in that wonderful wood-burning stove with warm crackers on top. Still, she kept that sense of humor. One time my brother Johnny and I were in the old barn with her while she was milking a cow. I was playing across from Grandma and I felt this wet, warm stream running down the side of my cheek. I put my hand up to wipe it away and I looked at the direction it came from and there was Grandma, sitting on the stool, milking away never missing a beat, but I did notice her shoulders shaking up and down as she had a good laugh. I must say, she was a good shot with that cow udder. If she was here today, I'd say, "Great shot, Gramma, I want you on my team."

I remember the fresh smells of the farm from the grass and the sawmill, the cow bells' mournful sound at night, and the rooster's early morning wake-up call, the smell of Grandma's bread baking and apple pies. There was nothing better than her warm bread with her fresh churned butter melting into it. Those picnics we had with all the good food. When we all got together, everyone enjoyed the good food and we did eat. Oh boy did we eat! A lot of it was healthy food fresh from their garden.

I loved to visit the farm when all the family was there. All of my uncles and aunts were special to me. To go with Wilbur to collect the sap from the maple trees for that wonderful syrup was a treat and shucking corn in the corn crib with Ejay for the chickens was fun work but hard on the thumbs. Ejay and Floyd used to entertain us with their trumpet playing. Aunt Mary was sweet, the most like grandma, a hard worker. My mother was eager to help everyone, physically and verbally. She is a good mother still at age ninety-five. Aunt Helen became a nurse and was very much needed and always cared well for me, and Aunt Elaine was just the sweetest person God ever put on this earth. She was always happy and singing. Did she ever grow up? I can't say enough about her goodness, and everyone who knew her knows what I mean. She visited us one time in Florida, and Mother and I took her to St. Augustine. She was like a kid in a candy store. She didn't want anything for herself but was so excited to see everything. What a delight she was to be with, a real gem. Aunt Betty is the sophisticated one in the family and she too took good care of me many times through the years. Paul was the most like Grandpa and went even farther by becoming a full-time preacher. He married Violet, a most fun gal to be around, and I believe she put that twinkle in Paul's eyes. And her sister Dorthy, who is smart as a whip, married Uncle Merle who I dearly loved. He wrote me a letter every Christmas that he was alive and I so enjoyed his news. They had a wonderful garden that they shared with everyone who they met. Uncle Floyd is the one I have bragged about the most. He is famous because he developed a new and improved Saran Wrap while he was employed with Dow. He is now married to Shirley and she is another true gem. When she is talking with me, I always feel she is very interested in what I say. I know she and Uncle Floyd have intelligent, stimulating conversations. And E. Jay is famous for his running races all his life. He and Jean's son, Garry, look like Jack and my son, Mike. Jean is the most all-together person and very genuine. I like her a lot. I truly love all my family from Eph's farm. Even though most of us have gone off in all directions, the memories live on and for me they are good memories and worth the retelling.

Chapter 17

My Second Farm

Family photograph of author after a hunting success

I spent the first eighteen years of my life on my dad's farm and the next eight years in college. My first job in 1953 was working for Dow Chemical Company in Midland, Michigan. Our daughter Beverly was born in 1957 and at this time Jenny's mother came to live with us. She had lost her sight. She lived with us about ten years until she went to a nursing home. When Cindy was born in 1965, we moved into a larger four-bedroom house. Eric was born in 1966. In 1969, I bought ninety-seven acres of wild land, located eleven miles north of Midland, for bow hunting and as an investment.

In 1976 I was transferred to Granville, Ohio where we duplicated the house we built in 1964 in Midland. In 1981, I was divorced. This was the worst time of my life. Beverly was twenty-four, Cynthia was seventeen and Floyd Eric was fifteen. Not long after this I bought a small 105-acre farm, located seven miles north of Newark, Ohio.

The next sixteen years I lived alone on this farm. In 1986 I retired from Dow Chemical Company. However, I continued my professional career working for Findley Industries in Johnstown, Ohio. At the same time, I had more than I could handle alone with work on the farm.

I rented some of the farm to a neighbor. He grew corn and soybeans in the front fields, and the fields behind the hill were used to pasture his cattle. He also rented two barns. I installed about two miles of electric fence around the fields used for pasture. At this time, I made a list of things that I felt needed done.

- Maintain electric fence
- Cut firewood for my wood-burning stove
- Paint barn roofs
- Mow grass
- Landscape lawn
- Housework (wash clothing, cook, sweep, clean house)
- Maintenance work on house and barn
- Plant a small orchard (apple trees and various berries)
- Plant a small vegetable garden.

I planted twenty apple trees and bought a sprayer to attach to my Kubota tractor. The apple trees grew fine and I got lots of apples for the deer. However I could not control apple worms using a low toxicity spray from the local stores.

I planted one acre of red raspberries, black raspberries, blackberries and strawberries, and one acre of blueberries. The big problem with all of the berries was lack of rain during the berry season. It was

a lot of work watering all the berry plants with lawn sprinklers. However, I usually got good yields of berries. The first day that I opened
the black raspberry field for people to pick, I collected over $700.
These berries were very much in demand. I sold the berries already
picked or "pick your own."

The strawberry plants lasted about three years. At planting time,
I had chosen varieties that were advertised to be select. The following two years after I stopped selling strawberries, people stopped
me in the Granville Market and many called who wanted the sweet
strawberries.

After about ten years, the raspberry plants died out, mostly from
virus. Only blueberries remain.

My vegetable garden was always very productive. I saved wood
ashes from the wood-burning stove. Usually I filled one 55-gallon
barrel with wood ashes. Next I pulverized dry cow manure with a
small mulcher. Then the wood ashes and cow manure were mixed
and used for fertilizing the garden.

To plant sweet corn, I used the following steps.

1. Plow and disk the ground
2. Cut a ten-inch deep furrow
3. Sprinkle a heavy row of pulverized cow manure and a light
   amount of wood ashes
4. Mix the manure and wood ashes with a roto-tiller or with the
   furrowing plow
5. Plant corn seeds in the bottom of the row and cover with about
   one inch of soil.

After the corn grew to a height of about three inches, the space
between rows was cultivated. In this step, the furrow was partially
filled. This got rid of almost all of the weeds. Any weeds left were
removed with a hoe. This step was repeated when the corn stalks
were about twelve inches. This removed weeds and filled soil around
the plants.

My garden was very productive and I gave most of the vegetables
away. However, I was not accomplishing all the jobs I had on my list
and after sixteen years I grew tired of being alone. At this time I met
Shirley and we were married in 1997.

Living on the farm was a new experience for Shirley. Having lived there for sixteen years alone, it was a new experience for me too. The farm was a beautiful place to live. The big white house nestled next to the big hillside was about one-third mile off the highway. The lane ran between two fields with tall green corn or soybeans on both sides.

The first and most important job after our marriage was to remodel the house. We hired different people to do many of the improvements:

- Put in new windows throughout the house
- Rebuilt an enclosed side porch and it became a mud room/half bath
- Built a new wrap-around front porch
- Removed the shed attached to the back of the house, added family room and master bedroom with full basement
- Insulated walls
- Built area under wrap-around porch and repaired stone basement walls
- Put in new eave troughs and spouting
- Contract: new roof and new siding
- Put in new propane furnace and air conditioning
- Rebuilt kitchen, the former bathroom became a utility room and the former pantry and utility room became our breakfast room.

We found a piece of hand-hewed log in the crawl space under the kitchen and we found stone footing inside the existing basement wall. This had been the footing of a log cabin, the second cabin built on the farm.

The first cabin on this farm was built next to a spring which now runs into an open water trough located in front of the horse barn. The last Indian raid in Licking County occurred at that cabin in the late 1700s. At the time of the raid the mother hid her nine-year-old son in a small cabin with a trap door covered by a rug. The boy's life was spared, but both his parents were killed. Soon after the raid, a family traveling in a wagon from Fort Detroit to West Virginia found the boy. His family was buried in a small cemetery on a knoll on the west side of the farm, now known as the little "Coad" Cemetery, now with an iron fence and gate and with many of the headstones crumbling. This cemetery had been sold to the Coad family. I liked to visit this little cemetery from time to time and think of those who came first to this beautiful land.

Each spring wild purple violets lift up their beauty along the square, black wrought-iron fence and around the stones standing or crumbling, marking a history that can't be repeated.

The boy lived in West Virginia for ten years and then came back at age nineteen to claim his land. He built another log cabin, starting with stones dragged from the hillside. This cabin later became a kitchen. A big kitchen grew on the very spot where the early log cabin stood. Now we drew plans to remodel the house.

On the west side of the kitchen and the breakfast room, we placed windows so we could look west into the hillside, where we know the boy must have played. On the back of the house, we built a large family room and master bedroom with a full basement and with windows facing south, so we could look at a thick wall of trees where we could see turkeys and deer, especially in the winter when the leaves had fallen. Looking in all directions from the house, we enjoyed the changes of the seasons. We liked the yellow and scarlet fall colors of the sugar maples.

In the corner of the family room, we built a stone floor and stone walls extending six feet from the corners. Here we located a wood-burning stove. Each year I would seek out the best oak tree tops that would give the hottest fires in the cold winter months.

From the house, two lanes run diagonally up the steep hill. The first lane ran directly into the large stone quarry. This lane was long ago closed. Some large sugar maples stand in the center of the road. Likely this lane was not used since the late 1700s.

The second lane runs to the west to the top of the hill, along the west boundary. Here the lane divides with one lane running south along the west fence. The other lane runs east through the center of the property. The lanes meet on the east side and run north back down the hill to the field east of the house. The lanes provide access to the property for sightseeing and for hauling wood.

We drove our jeep to witness the landscape with each change in season.

A big water trough sits in the field in the southwest corner.

Sheep and cattle once grazed these grasses. We explored each area. In the early spring we counted the pink and white blossoms of the dogwood, hawthorns, crabapple, and cherry and some that we searched the books in order to identify. Dozens of wildflowers bloomed in spring, summer and fall. Fall comes with a splendid display of colors.

At the top of the hill we could look north over the top of our house across a wide expanse of farms and wooded lots. There were no houses near the back of our farm. It was a peaceful place to be alone and a wonderful place for visitors. The quietness was broken by the songs of many birds or the call of a lone turkey.

During the winter months we identified thirty species of birds at our feeders, which were placed outside our breakfast room window.

We enjoyed nine years living on my second farm. We now rent the house and fields and live in a condominium in Dayton, Ohio. The memory of fields and woods will linger on and on for as long as we live.

Floyd Romesberg's second farm

Floyd and Shirley Romesberg, 2009

Chapter 18

The Periodic Table

How far out of place? How unrelated? How can any information on the periodic table fit in the events which occurred in the story of Ephraim's farm?

The answer is in a dream. A dream which I thought for a long time was a real life event. This dream occurred when I was ten years old. The sun was above the tree line and the bright rays shone in the window to my little bedroom. I got up, dressed, and hurried downstairs. No one was in sight. What was I to do this day? Everyone must have gone to the Long Farm. I grabbed my straw hat and ran. Before I reached the field, I saw the paper lying under a greenbriar bush. It was a copy of the periodic table.

I first saw the table in that dream. I saw it next in a chemistry book in high school. It was then I decided to seek a career in chemistry. At that time I felt that there was something special to learn about the periodic table.

The periodic table is a tabular arrangement of the elements in rows and columns, highlighting the regular repetition of properties of the elements. The table lists the 109 known elements along with the composition of each. Except for hydrogen, all of the elements are made of the same three particles: protons, neutrons, and electrons.

The <u>proton</u> is located in the dense nucleus of the atom, it has a positive charge equal in magnitude, but opposite in sign to that of the electron, and a mass 1836 times that of the electron.

The <u>neutron</u> has a mass about equal to that of the proton, but has no electrical charge. It is located in the nucleus of the atom.

The <u>electron</u> is a very light particle that carries a unit negative charge and exists in the region around the positively charged nucleus.[14]

When I was four, I often went outside on a night when the stars shone with all their glory. I wondered, as many others have, what is a star? What makes the sun shine? From where did the stars, the sun, and the earth come? Also, what about plants and animals, and people?

Some of these questions have been answered, some have not been answered.

Scientists believe that stars were formed 13 billion years ago. The stars were formed from hydrogen gas. Volumes of hydrogen gas were compressed by gravitational forces and a nuclear reaction started by the high temperature. Hydrogen gas was converted into helium gas with the release of huge amounts of heat in the form of photons. These photons have mass too low to measure by any known means. However, the atom produced is lighter than the hydrogen atoms that combined to make the helium nucleus. There is a 7% loss in weight. This weight is lost in the form of energy, according to Einstein's equation:

$$E = MC^2$$

Where: E = Energy

M = Mass

C = Speed of light (3×10^8 m/sec)

And: $C^2 = 9 \times 10^{16} = 90,000,000,000,000,000$

The speed of light squared is a very big number. Thus, the heat generated and the temperatures reached in the nuclear reaction in the sun and stars are extremely high.

There are different types and sizes of stars. Some are much bigger and some much smaller than our sun.

An estimate of the occurrence of the elements in the crust of the earth to a depth of 24 miles, and in the atmosphere, has been made by L. W. Clark, and in meteorites by O. C. Farrington.[15]

Estimates of the Occurrence (%) of Elements in Earth's Crust,
Earth's Atmosphere and in Meteorites[15]

Element	Crust	Atmosphere	Meteorites
Argon		1.4	
Oxygen	47.3	23.0	10.1
Nitrogen		75.5	
Silica	27.7		5.2
Aluminum	7.9		0.4
Iron	4.5		72.1
Calcium	3.5		0.5
Sodium	2.5		0.2
Potassium	2.5		
Magnesium	2.2		3.8
Titanium	0.5		
Hydrogen	0.2		
Carbon	0.2		
Nickel			6.5
Sulfur			0.5
Cobalt			0.4
Phosphorus			0.1

About two-thirds of the terrestrial elements have been spectro-scopically detected in the atmosphere of the sun, and no new elements are present in the sun, stars or nebulae. The mean density of the earth is about 5.6 and hence the earth's core is supposed to consist largely of iron, with some nickel.[15]

Scientists believe that the stars burn at different rates, at different temperatures, at different pressures, and at different compositions of gases. Initially, pure helium is formed and eventually some or all the elements in the periodic table are formed. Eventually, the hydrogen is depleted, temperatures drop and the star loses stability and blows up. At that time, whatever elements were in the star are scattered over far distances. Sometime later new stars and planets like the earth form from new hydrogen and from the dust in the other parts of the exploded star or stars. This would explain why and how all the elements are distributed on the earth's crust.

Scientists have learned a great deal about the universe. They have made large telescopes. Hubble built the first and famous Hubble Telescope. With this telescope we can see stars in all directions, too many to count, numbers stretching into billions with more too faint to see. Yet from calculations involving gravity and other forces that hold the universe in equilibrium, scientists claim that 95% of the matter in the universe cannot be seen, and they call it "dark matter."

One remarkable feature about the universe is that all the elements are made in the suns from the same particles. For example, at a specific time a sun could have given helium and another time iron or carbon or even gold. Some day man may send out into space a robot that would sample the dust and rock orbiting a planet in search for gold or other rare elements.

Scientists have learned much about our solar system and the universe. The elements were made first. All of the elements could have started with hydrogen (or smaller elementary particle). However, no one knows where the starting material came from.

We know how plants grow from seed. We know that water and elements combine with carbon dioxide from the air and with energy from the sun to give carbohydrates, proteins, etc. But we don't know where the original plants came from.

Scientists know that about thirty of the elements in the periodic table are found in the human body. They understand how oxygen from the air is carried from the lungs in the blood to the cells to react with carbohydrates, protein, etc. to give energy and grow new cells. However, we do not know where life came from in the first place.

Great discoveries have been made in all branches of science and many advances will be made by scientists throughout the world in the future. The positive charged proton is the focal point of particle physics. Because of the positive charge, the proton can be projected into space and made to collide at high speed with other protons and other particles. Then the proton has been broken into sub-particles of specific characteristics. This likely will lead to a much greater scientific description of the elements than possible with the Periodic Table.

> *Gen 1-1 In the beginning God created the heaven and the earth.*
> *Gen 1-29 Then God said, "I give you every seed-bearing plant on the face of the earth, and every tree that has fruit with seed .... And every green plant for food.*
> *Gen 2-7 And the Lord God formed man from dust of the earth.*
> *Rev 10-7 But in the days when the seventh angel is about to sound his trumpet, the mystery of God will be accomplished.*
> *Dan 12-10 None of the wicked will understand, but those who are wise will understand.*

Epilogue
... And then there were more dreams ...

Ephraim's Dream—
A Monster Comes out from the Hillside

Pop enjoyed fishing. He often spoke about the trout they caught in Wilson Creek, using a wooden pole, string, a hook and worms. When he was about ten years old, he dreamed that, while he was fishing in Wilson Creek, he saw a big hole opening along the side of the nearby hill and a huge monster crawled out of the hole. About ten years later, the Somerset Coal Company dug a hole into the side of the hill and opened a coal mine. They laid steel dinkey tracks into the hole. Donkeys were used to pull the cars loaded with coal out to the tipple, where the cars were dumped into railroad cars. Later the donkeys were replaced by gasoline-powered motor cars.

Soon after the mine opened, about 1910, Pop started to cut posts (from his woods) which were used to prop up the ceilings of rock in the mine. One Saturday, many years later in 1937, I went along with my brother Merle to deliver a wagon load of posts to the Wilsoncreek Mine. While Merle unloaded and stacked the posts, I walked close to the pit opening. Suddenly I heard a deep rumbling noise. It was a foggy day, so that I could barely see the pit opening. Then I saw what looked like the head of a huge black snake. Two big eyes seemed to glisten in the fog. Coal black angular shapes ran down its back. Was this the monster in Pop's dream? As I stood beside the dinkey track, the "monster" rolled by. Then I could see that the big eyes were lights and the angular shapes were lumps of coal in the line of cars. I backed away as the "monster" rumbled by me. I looked down at my boots and saw that I was standing in a stream of water pouring out of the mine. I looked at the water more closely and saw dark, rust-colored stones underneath the water. The "monster" had disappeared into the tipple. I walked a little closer to the pit opening. I now could see a few feet into the mine. The toxic water kept flowing. The "monster," with the help of miners, had cut into a vein of water and layers of sulfur compounds. The water, diverted from its natural course, now poured over the sulfur compounds and became toxic. Within a short period of time after the mine had opened, this stream from the mine ran down to the creek and all the fish had died. Now, seventy years later, the toxic water still flows and there are no fish in Wilson Creek.

Floyd's Second Dream—Footsteps

After my divorce, I lived alone for sixteen years. Then I had a dream which was repeated with a few variations on four successive nights.

In this dream, I heard loud footsteps coming from my kitchen into my bedroom. Clomp! Clomp! Clomp! They seemed to circle my bed. Then a presence jumped up on my bed. I swung at it with my right hand. I hit nothing, spun and landed in a basket of clothing beside my bed. Then I awakened and wondered what this meant.

The next night, I had the very same dream, except that when I struck at "it," I landed on the floor. The third night as I struck out at it, I bumped my head on my nightstand. The fourth night as I lay dreaming, I again heard footsteps, only these were much lighter, softer. They stopped at the foot of my bed. There was a long silence and then a booming voice said "God!"

The next morning I got up early and went outside and watched a huge orange ball coming up in the east, a magnificent sunrise! I stood there and prayed that God would answer my prayer. About a week later, I met Shirley. We were married three months and eight days later.

Psalm138:7 Though I walk in the midst of trouble, You preserved my life, You stretch out your hand against the anger of my foes. With your right hand you save me.

I Peter:5-8, 9 Be Self-controlled and alert. Your enemy, the devil, prowls around like a roaring lion, looking for someone to devour. Resist him, standing firm in the faith, because you know that our brothers throughout the world are undergoing the same kind of suffering.

Floyd's Third Dream—Our Last Trip to the Long Farm

In our youth we had gone to the Long Farm many times to work and many times just to enjoy the beauty all around us. Also, squirrel hunting season offered ample time to observe all the trees, birds, deer and other small game.

During the twenty-four years I lived on the farm in Ohio, I spent as much time as I could in the wooded areas, the abandoned pasture lands, and the valley on the southwest corner of the farm. I often sat on an old cement watering trough there, reminiscing about the past. Little wonder that I would also dream about the experiences of my youth at night. I had such a dream in 1997 after I met Shirley. In the dream, my brother E. Jay and I were still boys and it was about our last trip to the Long Farm.

E. Jay and I left the house and I saw him climb the rail fence next to the red machine shed. I followed and we both sat on the top rail. "Where are we going?" E. Jay asked. "I don't know, but we have been here before. I seem to remember all those trees and over there is Uncle Milt's house," I answered.

Then it began to rain. The water came rushing down both sides of the road. I could hear the gurgling sounds as the water slid over the stones in the ditches. I saw a familiar figure, a little boy, stepping into the water as if testing to see if his red and black boots would keep the cold water from his toes.

The rain stopped and I looked at the rows of pear and apple trees. There were two bushel baskets piled high with golden yellow ripe pears, ready to be taken to the pantry for Mom to can. We saw Merle and Paul loading sacks of apples onto the wagon, ready for the apple butter factory.

Moments in time flashed by. I saw Elaine peeling potatoes in the kitchen, Wilbur driving the team of horses with a wagon load of hay, Betty and Mom milking the cows, Helen sweeping the floor, Mom pulling loaves of hot bread from the oven, and both Pop and Mom watering the garden together.

Suddenly we jumped off the fence and started running. We ran through the orchard, past Henry's grave, through Leskey's field, past the big chestnut tree where Pop had once shot four squirrels, past the greenbriar bush where I found the paper, and into the big field. There we stopped. There was no one in sight. Then we ran again. We ran the length of the field, all the way to where the old house once stood. We stopped to catch our breath.

We drank from the spring. E. Jay asked, "Where are we going?" I answered "I don't know but we better hurry. We're a mile from home and it will soon be dark." We ran until we reached Harrison Pritts' house.

Then we walked down the road and onto the path to our home, past the pine trees, past Bingo's grave and finally to the old wire fence. Then we saw "it."

Everything appeared to be golden: the wire, the grass, the trees, fence posts, rocks, gates and barn. We got down on our knees and touched the ground. Everything was golden! Everything except the house. It was still white.

Then as it was getting dark, a bright light shone from the back corner of the house. The sky was bright. The light from the corner of the house was like a great beam. Then there was a man standing in front of the porch door. His skin seemed bronze and he wore a band uniform. His jacket and hat were covered with gold medals. Then I cried out, "E. Jay, E. Jay! We're in heaven. Look, there is Pop!"

The Gold – *Revelations 20:18 "The wall was made of jasper and the city of pure gold."*

The Light – *Revelations 21:23 "The city does not need the sun or the moon to shine on it, for the glory of God gives it light, and the Lamb is its lamp."*

The Gold Medals – *1 Peter 5:4 "And when the Chief Shepherd appears, you will receive the crown of glory that will never fade away."*

Revelations 2:10 "Be faithful, even to the point of death, and I will give you the crown of life."

II Corinthians 5:10 "*For we must all appear before the judgment seat of Christ, that each one may receive what is due him for the things done in the body, whether good or bad.*"

II Timothy 4:8 – "*Now, there is in store for me the crown of righteousness, which the Lord, the righteous Judge will award to me on that day, and not only to me, but also to all who have longed for his appearing.*"

Daniel 12:9 – "*Those who are wise will shine like the brightness of the heaven, and those who lead many to righteousness, like the stars forever and ever.*"

Endnotes

1 *Romesberg Family History,* Jack Romesberg, Dwight Romesberg, Roy and Ruth Romesberg, Minnie Sechler Romesberg, Pearl Romesberg, Lynn Romesberg, Abraham Romesberg.

2 *World Book Encyclopedia,* pp. 7-19, Scott Fietzer Co.

3 *Britannica Encyclopedia,* 15th Edition, VII, pp. 7-11.

4 *Pennsylvania* by Douglas Root, Fodor's Travel Publications, Inc.

5 Correspondence with Austin Romesberg, Ken and Betty Clarke.

6 *Genealogy of the Printz, Pritz, Pritts Family of Western Pennsylvania,* by Irvin Pritts.

7 *Wilsoncreek,* Irvin Pritts, http:/familytreemaker.geneology.com/users/p/v/i/Irvin-M-Prittts/file/0007/page.html.

8 *Generation Upon Generation,* Somerset County Bicentennial (1795-1995), Supplement to the *Daily American,* April 17, 1995, p. 38.

9 *Down the Road of Our Past,* Book 2, Rockwood Area Historical & Genealogical Society.

10 *Black Township Echoes,* Vol 1, No. 1, July 4, 1976.

11 *Down the Road of Our Past,* Book 3, Rockwood Area Historical & Genealogical Society.

12 *Down the Road of Our Past,* Book 1, Rockwood Area Historical & Genealogical Society.

13 *The History of Bedford, Somerset, Fullton Counties, Pennsylvania,* 1884, Early Pioneers.

14 From notes taken in chemistry classes at Penn State University and at the University of Cincinnati.

15 *General and Inorganic Chemistry for University Students,* T.R. Partington, Macmillian and Co. 1949.

About the Author

After graduating from Rockwood High School, Floyd went on to college, graduating from Penn State with a B.S. degree in Chemical Engineering. He received his M.S. degree at Bucknell and his Ph.D. in Chemistry from the University of Cincinnati in 1953. He was employed by The Dow Chemical Company in Midland, Michigan and transferred to their Granville, Ohio facility in 1975. While working for Dow, he was involved in industrial research, conducting process development and product improvement of fabricated plastic products.

He joined "Toastmasters International" for self-improvement in both speech and writing. He was active in this group for eight years.

After retiring from Dow in 1986, he worked for several companies, doing research and consulting. He finally retired altogether at the age of 77.

His avocation has been gardening extensively, both blueberries and vegetables. He also has had a life-long passion for hunting.

He now lives in Dayton, Ohio with Shirley, his wife of eleven years.

Floyd E. Romesberg, Ph.D.
February 15, 2009

Made in the USA
Lexington, KY
01 January 2015